Handbook on Design of Piles and Drilled Shafts Under Lateral Load

U. S. Department of Transportation
Federal Highway Administration

University Press of the Pacific
Honolulu, Hawaii

Handbook on Design of Piles and Drilled Shafts
Under Lateral Load

by
Lymon C. Reese

for Federal Highway Administration

ISBN: 1-4102-2560-7

Copyright © 2006 by University Press of the Pacific

Reprinted from the 1984 edition

University Press of the Pacific
Honolulu, Hawaii
http://www.universitypressofthepacific.com

All rights reserved, including the right to reproduce
this book, or portions thereof, in any form.

FOREWORD

This handbook was prepared for the Federal Highway Administration course on "Design of Piles and Drilled Shafts under Lateral Load. It was designed for use by geotechnical and structural engineers involved in design of foundations for highway structures.

Several methods of analysis and design of piles under lateral loading are presented. These include a computerized method of analysis using the program COM 624, a nondimensional solution, and the Broms method. In addition, analysis of pile groups, and method of testing piles under lateral load are included. Step-by-step procedures including example problems for each method are presented to help engineers to better understand the methods shown in the handbook.

R.J. Betsold
Director, Office of Implementation

PREFACE

Six years ago the Office of Implementation of the Federal Highway Administration sponsored the writer and colleagues in the preparation of a two-volume work entitled, "Design Manual for Drilled Shafts." Volume 2 of that work presented design procedures for drilled shafts that are subjected to lateral loads. This present volume, a handbook for highway engineers, is related to piles and drilled shafts and up-dates and enlarges the 1977 work.

The information presented herein is timely for several reasons; among them are that computational ability is greater now than ever before and is growing, technology on deep foundations under lateral load allows designs to be made with reasonable confidence, and the cost of construction encourages more sophisticated engineering. The use of a digital computer is emphasized, and details are given in this manuscript on the use of the computer program COM624. However, the verification of the results from the computer is strongly encouraged and considerable attention is given to techniques for such verification.

While the material given herein constitutes the best approach to design currently available, research is continuing on the behavior of piles and pile groups. The users are encouraged to stay abreast of such research and to make use of pertinent results.

A companion work is under preparation for the Office of Research of FHWA and is entitled, <u>Behavior of Piles and Pile Groups under Lateral Load</u>. That work gives more detail than is given here and the reader is encouraged to make use of the manual as need arises.

A draft of this handbook was used in two, two-day workshops, one in Austin, Texas, on June 29-30, 1983, and one in Albany, New York, on July 20-21, 1983.

Austin, Texas
April, 1984

METRIC CONVERSION FACTORS

APPROXIMATE CONVERSIONS FROM METRIC MEASURES

SYMBOL	WHEN YOU KNOW	MULTIPLY BY	TO FIND	SYMBOL
		LENGTH		
in	inches	2.5	centimeters	cm
ft	feet	30	centimeters	cm
yd	yards	0.9	meters	m
mi	miles	1.6	kilometers	km
mm	millimeters	0.04	inches	in
cm	centimeters	0.4	inches	in
m	meters	3.3	feet	ft
m	meters	1.1	yards	yd
km	kilometers	0.6	miles	mi
		AREA		
sq in	square inches	6.5	square centimeters	cm^2
sq ft	square feet	0.09	square meters	m^2
sq yd	square yards	0.6	square meters	m^2
sq mi	square miles	2.6	square kilometers	km^2
	acres	0.4	hectares	ha
cm^2	square centimeters	0.16	square inches	sq in
m^2	square meters	1.2	square yards	sq yd
km^2	square kilometers	0.4	square miles	sq mi
ha	hectares (10,000 m^2)	2.5	acres	
		MASS (weight)		
oz	ounces	28	grams	g
lb	pounds	0.45	kilograms	kg
	short tons (2000 lb)	0.9	tonnes	t
g	grams	0.035	ounces	oz
kg	kilograms	2.2	pounds	lb
t	tonnes (1000 kg)	1.1	short tons	

APPROXIMATE CONVERSIONS FROM METRIC MEASURES, continued

SYMBOL	WHEN YOU KNOW	MULTIPLY BY	TO FIND	SYMBOL
		VOLUME		
tsp	teaspoons	5	millileters	ml
tbsp	tablespoons	15	millileters	ml
fl oz	fluid ounces	30	millileters	ml
c	cups	0.24	liters	ℓ
pt	pints	0.47	liters	ℓ
qt	quarts	0.95	liters	ℓ
gal	gallons	3.8	liters	ℓ
cu ft	cubic feet	0.03	cubic meters	m^3
cu yd	cubic yards	0.76	cubic meters	m^3
ml	millileters	8.03	fluid ounces	fl oz
ℓ	liters	2.1	pints	pt
ℓ	liters	1.06	quarts	qt
ℓ	liters	0.26	gallons	gal
m^3	cubic meters	36	cubic feet	cu ft
m^3	cubic meters	1.3	cubic yards	cu yd
		TEMPERATURE (exact)		
°F	Fahrenheit temperature	5/9 (after subtracting 32)	Celsius temperature	°C
°C	Celsius temperature	9/5 (then add 32)	Fahrenheit temperature	°F

TABLE OF CONTENTS

	Page
Preface	ii
Metric Conversion Factors	iii
List of Figures	xi
List of Tables	xvii
List of Notations	xix

		Page
Chapter 1.	Introduction	1
	Occurrence of Laterally Loaded Piles	1
	Single-Column Support for a Bridge	1
	Support for an Overhead Sign Structure	2
	Pile-Supported Bridge	2
	Pile-Supported Bridge Abutment	5
	Pile-Supported Structures for Protection of Piers	7
	Foundation for an Arch Bridge	7
	Desirability of Performing Engineering Analyses in Design	10
	Nature of the Problem	10
	Interaction of Soils and Piles	11
	Analysis of a Pile Group	14
	General Methods of Solution of the Isolated Pile	14
	Design Problems	14
	Factor of Safety	15
	Influence of Analytical Method on Engineering Practice	17
	Organization of Handbook	20
	Design Organization	20
Chapter 2.	Soil Response	23
	Nature of Soil Response	23
	Effects of Loading and Presence of Water	26
	Method of Treating a Layered Soil Profile	27
	p-y Curves for Soft Clay	28
	Field Experiments	28
	Recommendations for Computing p-y Curves	29
	Recommended Soil Tests	32
	Example Curves	32
	p-y Curves for Stiff Clay Below the Water Table	35
	Field Experiments	35
	Recommendations for Computing p-y Curves	35
	Recommended Soil Tests	41
	Example Curves	41

TABLE OF CONTENTS (continued)

	Page
p-y Curves for Stiff Clay Above the Water Table	45
Field Experiments	45
Recommendations for Computing p-y Curves	45
Recommended Soil Tests	48
Example Curves	48
p-y Curves for Clay Below Water Table by Unified Method	50
Introduction	50
Recommendations for Computing p-y Curves	51
Example Curves	56
p-y Curves for Sand	57
Field Experiments	58
Recommendations for Computing p-y Curves	59
Recommended Soil Tests	64
Example Curves	64
p-y Curves for Rock	67
Chapter 3. Computer Method of Analysis	71
Analytical Method	71
Boundary Conditions	73
Computer Program COM624	74
Step-by-Step Procedure of Analysis	74
Example Problem, Steel Pile Supporting Bridge Abutment	75
Pile	75
Soil	75
Application and Loading of Foundation for Bridge Abutment or for a Retaining Wall	75
Computation of Yield Moment	76
Design Approach	76
Input Data	76
Output of Computer Program	78
Results of Computer Solutions	78
Behavior of Pile Under Service Loads	78
Example Problem, Steel Pile Supporting Bridge Pier	81
Example Problem, Drilled Shaft Supporting Bridge Abutment	81
Drilled Shaft	81
Computation of Bending Stiffness and Ultimate Moment	81
Results from Computer Solutions	85
Chapter 4. Nondimensional Solutions	87
Variation of Soil Modulus with Depth	87
Pile Head Free to Rotate (Case I)	89
Pile Head Fixed Against Rotation (Case II)	92
Pile Head Restrained Against Rotation (Case III)	101

TABLE OF CONTENTS (continued)

	Page
Step-by-Step Solution for Example Problem 1	102
Pile, Soil, and Loading	102
p-y Curves	102
Step 1	102
Step 2	102
Step 3	103
Step 4	103
Find Value of Relative Stiffness Factor T	103
Trial 1	104
Trial 2	106
Interpolation for T	106
Computation of Deflection and Bending Moment	107
Comparison with Computer Solution	109
Step-by-Step Solution for Example Problem 2	109
Pile, Soil, and Loading	109
p-y Curves	111
Find Value of Relative Stiffness Factor T	111
Computation of Deflection and Bending Moment	112
Comparison with Computer Solution	118
Chapter 5. Broms Method	119
Concepts Employed in Method	119
Piles in Cohesive Soil	119
Assumed Soil Response	119
Free-Head Piles in Cohesive Soil	120
Short, Free-Head Piles in Cohesive Soil	120
Long, Free-Head Piles in Cohesive Soil	122
Influence of Pile Length, Free-Head Piles in Cohesive Soil	123
Fixed-Head Piles in Cohesive Soil	124
Short, Fixed-Head Piles in Cohesive Soil	124
Intermediate Length, Fixed-Head Piles in Cohesive Soil	124
Long, Fixed-Head Piles in Cohesive Soil	125
Influence of Pile Length, Fixed-Head Piles in Cohesive Soil	126
Deflection of Piles in Cohesive Soil	127
Effects of Nature of Loading on Piles in Cohesive Soil	130
Detailed Step-by-Step Procedure for Piles in Cohesive Soil	130
Example Solution for Piles in Cohesive Soil	132
Piles in Cohesionless Soil	133
Assumed Soil Response	133
Free-Head Piles in Cohesionless Soil	134
Short, Free-Head Piles in Cohesionless Soil	134

TABLE OF CONTENTS (continued)

	Page
Long, Free-Head Piles in Cohesionless Soil	136
Influence of Pile Length, Free-Head Piles in Cohesionless Soil	136
Fixed-Head Piles in Cohesionless Soil	138
Short, Fixed-Head Piles in Cohesionless Soil	138
Intermediate Length, Fixed-Head Piles in Cohesionless Soil	138
Long, Fixed-Head Piles in Cohesionless Soil	139
Influence of Pile Length, Fixed-Head Piles in Cohesionless Soil	139
Deflection of Piles in Cohesionless Soil	140
Effects of Nature of Loading on Piles in Cohesionless Soil	141
Detailed Step-by-Step Procedure for Piles in Cohesionless Soil	142
Example Solution for Piles in Cohesionless Soil	143
Example Solution of a Drilled Shaft	144
Solution	144
Chapter 6. Analysis of Pile Groups	147
Response to Lateral Loading of Pile Groups	147
Widely-Spaced Piles	147
Model of the Problem	147
Detailed Step-by-Step Solution Procedure	152
Example Problem	153
Closely-Spaced Piles	159
Efficiency Formulas	159
Single-Pile Method	160
Chapter 7. Field Tests of Piles	163
Introduction	163
Selection of Test Site	164
Investigation of Soil Properties	165
Selection of Test Pile	165
Installation of Test Pile	166
Testing Procedures	167
Testing Pile with No Internal Instrumentation	169
Preliminary Computations	170
Obtaining Stiffness of Test Pile	170
Pile Installation	170
Loading Arrangement	171
Instrumentation	174
Interpretation of Data	176
Example Computation	176

TABLE OF CONTENTS (continued)

	Page
Testing Pile with Internal Instrumentation	178
Preliminary Computations	178
Instrumentation	179
Analysis of Data and Correlations with Theory	179
Review of Experiments Using Piles with Internal Instrumentation	180
Concluding Comments	181

Appendix 1. Information on Input of Data for Computer Program COM624 — 183

Introductory Remarks	184
Preparation for Input	184
Line-By-Line Input Guide	188
Summary	197
Error Messages	209
Coding Form for COM624	210

Appendix 2. Input and Output of Computer Program COM624 for Example Problems — 217

Input of Example Problem in Section 3.5	218
Output of Example Problem in Section 3.5	223
Input of Example Problem in Section 3.6	229
Output of Example Problem in Section 3.6	234
Input of Example Problem in Section 3.7 for PMEIX	238
Output of Example Problem in Section 3.7 for PMEIX	240
Input of Example Problem in Section 3.7 for COM624	245
Output of Example Problem in Section 3.7 for COM624	250

Appendix 3. Computer Program PMEIX — 255

Input Guide	256
Example Problems	266
Listing of Program	267
Listing of Input for Example Problems	275
Output of Example Problems	281

Appendix 4. Forms for Making Solutions Using Nondimensional Method — 293

Nondimensional Analysis of Laterally Loaded Piles with Pile Head Restrained Against Rotation	294
Nondimensional Analysis of Laterally Loaded Piles with Pile Head Fixed Against Rotation	295
Nondimensional Analysis of Laterally Loaded Piles with Pile Head Free to Rotate	296

TABLE OF CONTENTS (continued)

	Page
Appendix 5. Order Form for Computer Programs COM624 and PMEIX	297
Appendix 6. Modification of p-y Curves for Battered Piles	299
Appendix 7. Example Problems for Workshops	303
Exercise Relating to Material Presented in Chapter 2	304
Exercise Relating to Material Presented in Chapter 3 (Case 1)	306
Exercise Relating to Material Presented in Chapter 3 (Case 2)	312
Exercise Relating to Material Presented in Chapter 3 (Case 3)	318
Exercise Relating to Material Presented in Chapter 4	324
Exercise Relating to Material Presented in Chapter 5	330
Appendix 8. ASTM Standards	333
References	355

LIST OF FIGURES

Figure No.	Title	Page
1	Single column supports	3
2	Loadings on single-column support for a bridge	3
3	Overhead sign	4
4	Elevation view of an overhead sign structure (a) two-pile foundation (b) single-pile foundation	4
5	Bridge supported by piles	5
6	Bridge abutment	6
7	Sketch of a pile-supported bridge abutment	6
8	Bridge with dolphin	8
9	Sketch of a pier-protective system (a) single pile employed as a dolphin (b) design concept	8
10	Arch bridge	9
11	Pile-supported foundation for an arch bridge	9
12	Soil-structure interaction for a strip footing	12
13	Model of a pile under axial load	12
14	Model of a pile under lateral load showing concept of soil response curves	13
15	Influence of pile penetration on groundline deflection	16
16	Retaining wall for example computation	19
17	Sign supports for example computations	21
18	Pipe pile and soil elements	23
19	Conceptual p-y curve	24
20	Wedge-type failure of surface soil	25
21	Potential failure surfaces generated by pile at several diameters below ground surface	26

LIST OF FIGURES (continued)

Figure No.	Title	Page
22	Characteristic shapes of the p-y curves for soft clay below the water table (a) for static loading (b) for cyclic loading (from Matlock, 1970)	29
23	Soil profile used for example p-y curve for soft clay	33
24	Example p-y curves for soft clay below water table, Matlock criteria, cyclic loading	33
25	Characteristic shape of p-y curve for static loading in stiff clay below the water table (after Reese, Cox, Koop, 1975)	36
26	Values of constants A_s and A_c	37
27	Characteristic shape of p-y curve for cyclic loading in stiff clay below water table (after Reese, Cox, Koop, 1975)	40
28	Soil profile used for example p-y curves for stiff clay	41
29	Example p-y curves for stiff clay below the water table, Reese criteria, cyclic loading	42
30a	Characteristic shape of p-y curve for static loading in stiff clay above water table	46
30b	Characteristic shape of p-y curve for cyclic loading in stiff clay above water table	47
31	Example p-y curves for stiff clay above water table, Welch criteria, cyclic loading	49
32	Characteristic shape of p-y curve for unified clay criteria - static loading	52
33	Characteristic shape of p-y curve for unified clay criteria - cyclic loading	56
34	Example p-y curves for soft clay below water table, unified criteria, cyclic loading	57
35	Example p-y curves for stiff clay below water table, unified criteria, cyclic loading	58

LIST OF FIGURES (continued)

Figure No.	Title	Page
36	Characteristic shape of a family of p-y curves for static and cyclic loading in sand	59
37	Values of coefficients $\bar{A}_c$ and $\bar{A}_s$	62
38	Nondimensional coefficient B for soil resistance versus depth	62
39	Example p-y curves for sand below water table, Reese criteria, cyclic loading	65
40	Recommended p-y curve for design of drilled shafts in vuggy limestone	69
41	Results from computer solutions for example problem, steel pile supporting bridge abutment	79
42	Bending moment versus depth for example problem, steel pile supporting bridge abutment	80
43	Deflection versus depth for example problem, steel pile supporting bridge abutment	80
44	Results from computer solutions for example problem, steel pile supporting bridge pier	82
45	Interaction diagram for drilled shaft of example problem, from Computer Program PMEIX	83
46	Results from computer solutions for example problem, drilled shaft supporting bridge abutment	86
47	Form of variation of soil modulus with depth	88
48	Pile deflection produced by lateral load at mudline	90
49	Pile deflection produced by moment applied at mudline	91
50	Slope of pile caused by lateral load at mudline	93
51	Slope of pile caused by moment applied at mudline	94
52	Bending moment produced by lateral load at mudline	95

xiii

LIST OF FIGURES (continued)

Figure No.	Title	Page
53	Bending moment produced by moment applied at mudline	96
54	Shear produced by lateral load at mudline	97
55	Shear produced by moment applied at mudline	98
56	Deflection of pile fixed against rotation at mudline	100
57	Computed p-y curves for Example Problem 1	105
58	Graphical solution for relative stiffness factor, Example Problem 1	107
59	Comparison of bending moment from computer and from nondimensional solution, Example Problem 1	110
60	Comparison of deflection from computer and from nondimensional solution, Example Problem 1	110
61	Pile, soil and loading for Example Problem 2	111
62	Trial plots of soil modulus values, Example Problem 2	116
63	Interpolation for final values of relative stiffness factor T, Example Problem 2	116
64	Comparison of deflection and bending moment from computer and from nondimensional solution, Example Problem 2	118
65	Assumed distribution of soil resistance for cohesive soil	119
66	Deflection, load, shear, and moment diagrams for a short pile in cohesive soil that is unrestrained against rotation	120
67	Design curves for short, free-head piles under lateral load in cohesive soils (after Broms)	121
68	Design curves for long piles under lateral load in cohesive soils (after Broms)	122

LIST OF FIGURES (continued)

Figure No.	Title	Page
69	Deflection, load, shear, and moment diagrams for an intermediate-length pile in cohesive soil that is fixed against rotation at its top	125
70	Failure mode of a short pile in cohesionless soil that is unrestrained against rotation	135
71	Deflection, load, shear, and moment diagrams for a short pile in cohesionless soil that is unrestrained against rotation	135
72	Design curves for long piles under lateral load in cohesionless soils (after Broms)	137
73	Typical pile-supported bent	148
74	Simplified structure showing coordinate systems and sign conventions (after Reese and Matlock, 1966) (a) with piles shown (b) with piles represented as springs	149
75	Set of pile resistance functions for a given pile (a) axial pile resistance versus axial displacement (b) lateral pile resistance versus lateral pile displacement (c) moment at pile head versus lateral pile displacement for various rotations (α_p) of the pile head	151
76	Sketch of a pile-supported retaining wall	154
77	Interaction diagram of reinforced concrete pile	155
78	Axial load versus settlement for reinforced concrete pile	155
79	Pile loading, Case 4	158
80	Plan and elevation of foundation analyzed in example problem	162
81	Simultaneous testing of two drilled shafts under lateral loading, Skyway Bridge, Tampa Bay, Florida	163
82	Two-pile test arrangement for two-way loading	172

List of Figures (continued)

Figure No.	Title	Page
83	Two-pile test arrangement with one-way loading	173
84	Schematic drawing of deflection-measuring system	174
85	Device for measuring pile-head rotation	175
86	Information for analysis of test at St. Gabriel	177
87	Comparison of measured and computed results for St. Gabriel test	177
A1.1	Coordinate system	185
A1.2	Sign convention	186
A1.3	Flowchart for input to program COM624	198
A3.1	Portion of a beam subjected to bending	258
A3.2	Beam cross-section for example problem	258
A3.3	Stress-strain curve for concrete used by program PMEIX	261
A3.4	Stress-strain curve for steel used by program PMEIX	261
A3.5	Data input form for Computer Program PMEIX	263
A3.6	Concrete column cross-sections for example problem	266
A6.1	Modification of p-y curves for battered piles (after Kubo, and Awoshika and Reese)	301

LIST OF TABLES

Table No.	Title	Page
1	Representative values of ε_{50}	30
2	Representative values of k for stiff clays	37
3	Representative values of ε_{50} for stiff clays	38
4	Representative values for ε_{50}	52
5	Curve parameters for the unified criteria	53
6	Representative values for k	54
7	Nondimensional coefficients for p-y curves for sand (after Fenske)	60
8	Relationship between N and ϕ and D_r (after Gibbs and Holtz, 1957)	63
9	Representative values of k (lb/cu in) for sand	64
10	Data for one of the computer runs	77
11	Computed values of bending stiffness for drilled shaft of example problem	84
12	Moment coefficients at top of pile for fixed-head case	99
13	Values of p versus y for soil pile system at a depth of 10 inches, Example Problem 1	103
14	Computed p-y curves for Example Problem 1	104
15	Values of E_s versus depth for trial 1, Example Problem 1	106
16	Values of E_s versus depth for trial 2, Example Problem 1	108
17	Values of deflection and bending moment along pile length for T = 45.6 inches, Example Problem 1	108
18	Computed p-y curves for Example Problem 2	112
19	Nondimensional analysis of laterally loaded piles with pile head free to rotate, trial 1 for Example Problem 2	113

LIST OF TABLES (continued)

Table No.	Title	Page
20	Nondimensional analysis of laterally loaded piles with pile head free to rotate, trial 2 for Example Problem 2	114
21	Nondimensional coefficients	115
22	Values of deflection and bending moment along pile length for T = 88 inches, Example Problem 2	117
23	Table of functions for pile of infinite length	128
24	Terzaghi's recommendations for soil modulus α for laterally loaded piles in stiff clay	129
25	Terzaghi's recommendations for values of k for laterally loaded piles in sand	141
26	Values of loading employed in analyses	156
27	Computed movements of origin of global coordinate system	157
28	Computed movements and loads at pile heads	157
A1.1	Summary of input variables and formats	202
A1.2	Dictionary - Input variables	203
A3.1	Detailed input guide with definitions of variables	264

LIST OF NOTATIONS

a_i	horizontal coordinate of global axis system in pile group analysis
a_y, a_s, a_m, a_v, a_p	nondimensional coefficients, same as A-coefficient except for rigid-pile theory
A	coefficient used to define the shape of the p-y curve, unified criteria for clay
A_c	empirical coefficient used in equations for p-y curves for stiff clays below water surface, cyclic loading
A_s	empirical coefficient used in equations for p-y curves for stiff clays below water surface, static loading
A_y, A_s, A_m, A_v, A_p	nondimensional coefficients in elastic-pile theory relating to an applied force P_t, for deflection, slope, moment, shear and soil reaction, respectively
$\bar{A}_c$	empirical coefficient used in equations for p-y curves for sand, cyclic loading
$\bar{A}_s$	empirical coefficient used in equations for p-y curves for sand, static loading
$A_0, A_1, \ldots A_m$	coefficients in solutions for the difference equation method
A_{1t}	deflection coefficient for long pile with pile top restrained against rotation
b	pile diameter (L)
b_i	vertical coordinate of global axis system in pile group analysis (L)
B	width of foundation (L)
B_c	empirical coefficient used in equations for p-y curves for sand, cyclic loading
B_s	empirical coefficient used in equations for p-y curves for sand, static loading
B_y, B_s, B_m, B_v, B_p	nondimensional coefficients in elastic-pile theory relating to an applied moment M_t for deflection, slope, moment, shear and soil reaction, respectively

LIST OF NOTATIONS (continued)

$B_0, B_1, \ldots B_m$	coefficients in solutions for the difference equation method
c	undrained shear strength (F/L^2)
c_a	average undrained strength of clay from ground surface to depth (F/L^2)
C	coefficient related to stress level used in p-y curves for stiff clay above water surface
$\bar{C}$	coefficient used in equations for p-y curves for sand
$C_0, C_1, \ldots C_m$	coefficients in solutions for the difference equation method
e	eccentricity of lateral load (L)
E	Young's modulus (F/L^2)
E	efficiency factor for pile group (1 or < 1)
E_{cp}	soil modulus after cyclic loading (F/L^2)
E_s	soil modulus (secant to p-y curve) (F/L^2)
E_{si} or $E_{s(max)}$	initial or maximum soil modulus (F/L^2)
EI	flexural rigidity of pile (F/L^2)
f	the distance from the groundline to the point of M_{max} for cohesionless soil (L)
f_b	bending stress on pile (F/L^2)
f'_c	compressive strength of concrete (F/L^2)
f_y	yield point of steel rebar (F/L^2)
F	coefficient used to define deterioration of soil resistance at large deformations, unified criteria for clay, static loading
F_{h_i}	horizontal component of force on any "i-th" pile (F)
F_{mt}	moment coefficient at top of pile for fixed-head case
F_p	force against a pile in clay from wedge of soil (F)

LIST OF NOTATIONS (continued)

F_p	rate coefficient (limited data suggest a range of from 0.05 - 0.3)
F_{pt}	force against a pile in sand from a wedge of soil (F)
F_{v_i}	vertical component of force on any "i-th" pile (F)
F_y	deflection coefficient for fixed head pile
g	the distance from M_{max} to pile tip (L)
Δh	horizontal translation in global coordinate (L)
I	moment of inertia (L^4)
J	factor used in equation for ultimate soil resistance near ground surface for soft clay
J_m	= M_t/y_t modulus for computing M_t from y_t (F)
J_x	= P_x/x_t modulus for computing P_x from x_t (F/L)
J_y	= P_t/y_t modulus for computing P_t from y_t (F/L)
k	constant giving variation of soil modulus with depth (F/L^3)
k_c	coefficient used in equations for p-y curves for stiff clays below water surface, cyclic loading (F/L^3)
k_m	soil modulus at point m (F/L^3)
k_s	initial slope of p-y curve for sand (F/L^3)
$k_0, k_1, k_2, \ldots$	constants of soil modulus variation in $E_s = k_0 + k_1 X + k_2 X^2$
k_θ	= M_t/S_t, spring stiffness of restrained pile head (F-L)
K	= M_t/S_t, rotational restraint of pile top (F-L)
K_a	minimum coefficient of active earth pressure
K_o	coefficient of earth pressure at rest
K_p	Rankine coefficient of passive pressure
L	length of pile (L)

LIST OF NOTATIONS (continued)

LI	liquidity index for clay
m	pile node number
m	slope used in defining portion of p-y curve for sand
M	bending moment (F-L)
M_a	an applied moment at the top of the pile adopted in Broms' method (F-L)
M_m	bending moment at node m (F-L)
M_{max}^{pos}	maximum bending moment in pile (F-L)
M_{max}	maximum positive bending moment (F-L)
M_{pos}, $M+$	positive moment (F-L)
M_{s_i}	bending moment at pile head of any "i-th" pile in pile group (F-L)
M_t	bending moment at pile head (F-L)
M_t/S_t	rotational restraint constant at pile top (F-L)
M_y, M_{yield}	yield moment of pile (F-L)
M_y^+, M_y^-	positive yield moment and negative yield moment of pile (F-L)
n	exponent used in equations for p-y curves for sand
N	number of cycles of load application used in p-y curves for stiff clay above water surface
O_R	overconsolidation ratio for clay
p	soil resistance (F/L)
P_c	ultimate soil resistance for pile in stiff clay below water surface (F/L)
P_{cd}	ultimate soil resistance at depth for pile in stiff clay below water surface (F/L)
P_{cR}	residual resistance on cyclic p-y curves, unified criteria for clay (F/L)

xxii

LIST OF NOTATIONS (continued)

P_{ct}	ultimate soil resistance near ground surface for pile in stiff clay below water (F/L)
P_h	total horizontal load on pile group (F)
P_k	a specific resistance on p-y curves for sand (F/L)
P_r	residual resistance on static p-y curves, unified criteria for clay (F/L)
P_s	ultimate soil resistance for pile in sand (F/L)
P_{sd}	ultimate soil resistance at depth for pile in sand sand (F/L)
P_v	total vertical load on pile group (F)
P_x	axial load at pile head (F)
P_{st}	ultimate soil resistance near ground surface for pile in sand (F/L)
P_t	lateral force at pile top (F)
P_u	ultimate soil resistance or ultimate soil reaction (F/L)
$(P_u)_{ca}$	ultimate soil resistance near ground surface for pile in clay (F/L)
$(P_u)_{cb}$	ultimate soil resistance at depth for pile in clay (F/L)
$(P_u)_{sa}$	ultimate soil resistance near ground surface for pile in sand (F/L)
$(P_u)_{sb}$	ultimate soil resistance at depth for pile in sand (F/L)
P_x	axial load at pile top (F)
P_{ult}	ultimate lateral load on a pile (F)
PI	plasticity index for clay
q_u	unconfined compressive strength of clay (F/L²)
$(Q_{ult})_G$	ultimate axial capacity of the group pile (F)
$(Q_{ult})_P$	ultimate axial capacity of an individual pile (F)

LIST OF NOTATIONS (continued)

R	relative stiffness factor
R_t	$= E_t I_t$, flexural rigidity at pile top (F-L²)
s_t	sensitivity of clay
S	slope (L/L)
S_t	rotation at groundline
T	$= \sqrt[5]{\frac{EI}{k}}$, relative stiffness factor in nondimensional solution method
v	shear (F)
W	distributed load along the length of the pile (F/L)
W_L	liquid limit for clay
x	coordinate along pile, beam (L)
x_r	transition depth at intersection of equations for computing ultimate soil resistance against a pile in clay (L)
x_t	transition depth at intersection of equations for computing ultimate soil resistance against a pile in sand (L)
x_t	vertical displacement at pile head (L)
y	pile deflection and for y-coordinate (L)
y_c	deflection coordinate for p-y curves for stiff clay above water surface, cyclic loading (L)
y_F	pile deflection with pile head fixed against rotation (L)
y_g	deflection at intersection of the initial linear portion and the curved portion of the p-y curve, unified criteria for clay (L)
y_k	a specific deflection on p-y curves for sand (L)
y_m	pile deflection at node m (L)
y_m	a specific deflection on p-y curves for sand (L)

LIST OF NOTATIONS (continued)

y_p	a specific deflection on p-y curves for stiff clay below water surface, cyclic loading (L)
y_s	deflection coordinate for p-y curves for stiff clay above water table, static loading (L)
y_t	pile top deflection (L)
y_u	a specific deflection on p-y curves for sand (L)
y_{50}	a specific deflection on p-y curves for clay (L)
Y_A, S_A, M_A, V_A, P_A	components of pile response due to an applied force P_t, namely, deflection, slope, moment, shear and soil reaction, respectively
Y_B, S_B, M_B, V_B, P_B	components of pile response due to an applied moment M_t, namely deflection, slope, moment, shear and soil reaction, respectively
Z	plastic modulus (L³)
Z	$= \frac{x}{T}$, depth coefficient in elastic-pile theory
Z_{max}	$= \frac{L}{T}$, maximum value of elastic-pile theory coefficient
α	$= E_s$, soil modulus, coefficient of subgrade reaction (F/L²)
α	angle used in defining geometry of soil wedge
α_s	rotational angle in the global coordinate system
α_T	Terzaghi's soil modulus for stiff clay (F/L²)
α_{pFkj}	the coefficient to get the influence of pile j on pile k
β	$\sqrt[4]{\frac{\alpha}{EI}}$, relative stiffness factor (1/L)
β	angle used in defining geometry of soil wedge
γ	average unit weight of soil (F/L³)
γ'	bouyant unit weight or average unit weight used in computing effective stress (F/L³)
ε	axial strain of soil
ε_{50}	axial strain of soil corresponding to one-half the maximum principal stress difference

LIST OF NOTATIONS (continued)

θ	angle of rotation
θ_i	the inclined angle between vertical line and pile axis of the "i-th" batter pile
μ	reduction factor used in pile groups analysis suggested by Japanese Road Association
ν	Poisson's ratio
σ_Δ	deviator stress (F/L²)
σ_v	average effective stress (F/L²)
ϕ	angle of internal friction of sand
$\phi(Z)$	$= E_s T^4/EI$, nondimensional soil modulus function of elastic pile theory

CHAPTER 1. INTRODUCTION

1.1 OCCURRENCE OF LATERALLY LOADED PILES

The principal use of piles in highway structures is for the support of bridges, either to span water courses or to serve in interchanges. However, there are other applications of piles: foundations for retaining walls, bridge abutments, and overhead signs. Piles may also be used for dolphins and to stabilize slopes.

The drilled shaft is really a pile that is constructed by making an excavation with an auger or a drilling bucket, placing a rebar cage if desirable, and filling the excavation with fluid concrete. Drilled shafts have been used to make a retaining structure when constructing an underpass. The drilled shafts are installed prior to starting the excavation for the underpass on lines at each side of the underpass. For convenience in installation, every other drilled shaft is installed and then the intermediate ones are drilled and placed. The excavation for the underpass is made with tiebacks being installed if the depth of excavation of the drilled shafts cannot support the earth pressures by cantilever action. For simplicity of expression, the term "pile" will be employed when referring to both driven piles and drilled shafts.

The lateral loads on the piles for the above structures are derived from earth pressures, centrifugal forces from moving vehicles, braking forces, wind pressures, current forces from flowing water, wave forces in some unusual instances, earthquakes, and impact loads from barges or other vessels. Even if none of the above sources of lateral loading are present, an analysis may be necessary to investigate the lateral deflection and bending moment that would result from the eccentric application of axial load.

There are numerous kinds of structures employed in highway construction where piles are subjected to lateral loading. Examples of some of these structures are presented in the following paragraphs, along with photographs and sketches, and some general comments are made about analytical techniques.

Single-Column Support for a Bridge

Figure 1 is a photograph showing single columns supporting bridge spans. The use of a single column is desirable when space is restricted or perhaps to give a better appearance to a structure being built in an urban area. Figure 2 is a sketch showing the potential loadings on the pile and pile cap. The pile in this case is assumed to be a drilled shaft that is continuous with the column.

The equations of statics can be employed to compute an axial load, a lateral load, and a moment (all lying in the same plane) at the groundline. The methods outlined herein can then be employed to compute the deflection and rotation at the groundline and the bending moment as a function of depth. Alternatively, the loadings can be computed at the base of the cap, and the computer program that is presented can be employed to analyze the column as well as the pile. The weight of the column itself would probably be neglected in the latter analysis.

Support for an Overhead Sign Structure

A photograph of an overhead-sign structure is shown in Fig. 3. Large numbers of such structures are required for the highway system. The principal loading on such a structure is derived from storm winds. The winds as a rule will be gusty, imposing a cyclic lateral load on the structure.

Figure 4 shows views of two types of foundation used for sign structures. Figure 4a shows a two-pile foundation. The shear, overturning moment, and vertical load can be computed at the pile cap and the foundation can be analyzed by the methods outlined in Chapter 6. Figure 4b shows a single-pile support that can be analyzed directly by the computer program described herein. It should be noted that the two-pile system resists the wind moment primarily by added tension and compression in the piles. The single-pile foundation resists moment by bending. The two foundation systems can be compared and the most favorable system can be selected.

Pile-Supported Bridge

A photograph of a bridge that is supported by piles is shown in Fig. 5. The forces that must be sustained by the piles include live loads, dead loads, wind loads, loads from a possible stream or reservoir (current forces and wave forces), centrifugal forces, and braking forces. Experiences in the field have shown that braking forces from one or more large vehicles can be sizeable, especially if heavily loaded trucks are suddenly brought to a stop on a downward-sloping approach span.

No sketches are presented to show the possible configurations of the piles supporting a bridge bent. An extremely large number of arrangements

Fig. 1. Single-column supports

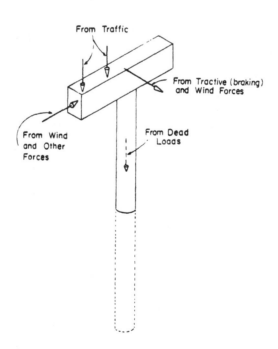

Fig. 2. Loadings on single-column support for a bridge

Fig. 3. Overhead sign

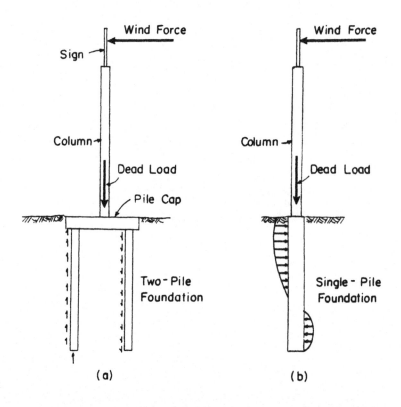

Fig. 4. Elevation view of an overhead sign structure (a) two-pile foundation and (b) single-pile foundation

Fig. 5. Bridge supported by piles

are possible with battered piles used frequently. The methods presented herein are applicable to the design of most systems.

An exception should be noted, however. If braking forces, acting longitudinally, occur simultaneously with forces from wind and current, acting transversely, the result will be biaxial bending. The methods described herein can be employed but some judgement must be employed by the designer in the summation of the computed stresses.

Pile-Supported Bridge Abutment

Figure 6 is a photograph of a bridge abutment. The assumption is made that the slope has been graded and that the piles supporting the abutment are then installed. A sketch of the abutment is shown in Fig. 7.

The lateral loadings that must be sustained by the piles can come from a downward creep of the slope or from braking forces that are transmitted through the deck system. If such loadings can be ascertained, the lateral and axial loadings on the piles can be used in the analysis of the

Fig. 6. Bridge abutment

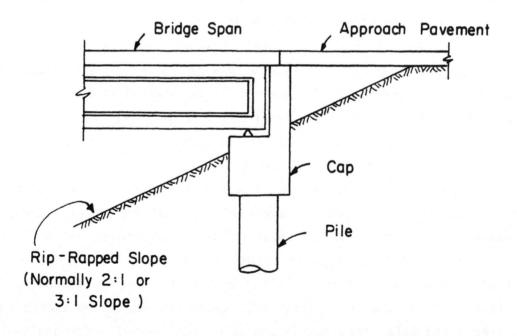

Fig. 7. Sketch of a pile-supported bridge abutment

piles. The computer program that is presented includes a feature to allow for distributed loading along a pile.

Pile-Supported Structures for Protection of Piers

There are many cases in which bridges must be constructed across a navigable stream or canal. Figure 8 presents a photograph of such a case. An important problem involves the design of a system to protect the bridge piers. The methods presented herein provide one means of designing such a protective system.

Figure 9a is a sketch of a possible protective system for a bridge pier. A dolphin consisting of a single pile is shown between the vessel and the pier. The design of the protective system must be based on assumptions concerning the mass of the vessel and its velocity at contact with the dolphin. Figure 9b shows the design concept that is employed. The single free-standing pile is contacted by the vessel and deflects until the kinetic energy in the vessel is balanced by the energy absorbed by the dolphin, as shown. The methods presented herein can readily be employed to compute a load-deflection curve, such as shown in Fig. 9b.

Rather than employing a single pile, multiple piles can be used, of course. They must be arranged in such a way that a sufficient number of them are deflected to absorb the energy of the design vessel and in consideration of the direction of approach of the design vessel. The use of free-standing piles as energy-absorbing structures is proven. Such structures are in use in many places. Driving the piles closely together and tying the piles together such that deflection is restricted should be avoided. Such structures become stiff and less absorbent of energy.

Foundation for an Arch Bridge

A photograph of an arch bridge is shown in Fig. 10. In some cases it might be desirable to employ piles in the foundations for the arch. Such a scheme is shown by the sketch in Fig. 11.

The problem of the design of the foundation involves the solution of numerous structural details and constitute the overriding effort. However, the methods presented in Chapter 6 should prove useful in the analysis of the large diameter drilled shafts. Because of the massive size of such a foundation, it will probably be necessary to consider that the behavior of the piles and cap will not follow exactly the elementary equations for bending that are derived from the elastic line.

Fig. 8. Bridge with dolphin

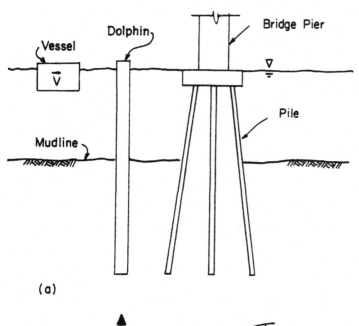

Fig. 9. Sketch of a pier-protective system
(a) single pile employed as a dolphin
(b) design concept

Fig. 10. Arch bridge

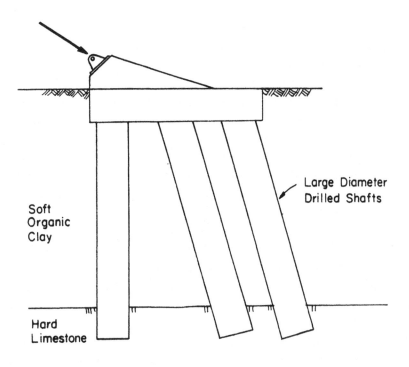

Fig. 11. Pile-supported foundation for an arch bridge

1.2 DESIRABILITY OF PERFORMING ENGINEERING ANALYSES IN DESIGN

Piles that must sustain lateral load can be designed, and have been designed successfully in many instances, by approximate methods. The allowable lateral load on a vertical pile can be obtained from a table of presumptive values as found in some handbooks or building codes. However, these allowable loads as a rule are small compared to values that may be computed by methods recommended herein. Another approach that has been used extensively is to assume that lateral loads are sustained by the lateral component of axial loads taken by piles that are installed at an angle with respect to the vertical (batter piles). Most methods that are available for the analysis of a pile group that includes batter piles are approximate in that the movements of the pile heads under load are not considered.

Experience has shown that a more rational and a more satisfactory solution of the problem of the pile under lateral loading is obtained by using the methods described herein rather than by using the approximate methods mentioned above. The computer program that is described allows computations to be made with little effort. The investment of engineering effort is compensated many times over by savings in labor and materials and by achieving a safer design in some cases.

1.3 NATURE OF THE PROBLEM

The application of a lateral load to the top of a pile must result in the lateral deflection of the pile. The reactions that are generated in the soil must be such that the equations of static equilibrium are satisfied, and the reactions must be consistent with the deflections. Also, because no pile is completely rigid, the amount of pile bending must be consistent with the soil reactions and the pile stiffness.

Thus, the problem of the laterally loaded pile is a "soil-structure-interaction" problem. The solution of the problem requires that numerical relationships between pile deflection and soil reaction be known and that these relationships be considered in obtaining the deflected shape of the pile.

The problem seems formidable; however, two technological advances have allowed the problem to be solved with relative ease. Instrumentation that enables strains to be read remotely has made possible the determination of soil response during the testing of full-scale piles. And the digital computer allows the deflected shape (lateral deflection) of a pile to be computed rapidly and accurately even though the soil reaction against the pile is a nonlinear function of pile deflection and depth below the ground surface.

The use of the computer method will be presented in the following sections. Some approximate methods of solution will also be given; these can be used to learn whether or not there are gross errors in the computer output. Further, the making of hand computations with the approximate methods serves to illustrate in detail the nature of the problem of the laterally loaded pile.

1.4. INTERACTION OF SOILS AND PILES

The nature of the soil-structure-interaction problem can be illustrated by considering the behavior of a strip footing, as shown in Fig. 12. The assumption is usually made that the bearing stress is uniform across the base of the footing as shown in the figure. However, under the stress distribution that is shown, the cantilever portion of the footing will deflect such that the downward movement at b is less than the downward movement at a. The footing is probably stiff enough that the deflection of b with respect to a is small; however, the concept is established that the base of the footing does not remain planar. Therefore, the bearing stress across the base of the footing conceptually should not be uniform.

While it has been shown that the behavior of a strip footing involves soil-structure interaction, little economic advantage is to be gained by taking the curvature of the footing into account. Such is not the case for a pile, however. Figure 13 shows a model of an axially loaded pile with the soil replaced by a set of mechanisms. The mechanisms show that the load transfer in skin friction and in end bearing are nonlinear functions of the downward movement of the pile. If the mechanisms can be described numerically, a nonlinear curve showing axial load versus pile-head movement can easily be obtained (Reese, 1978).

The two examples of soil-structure interaction given thus far serve to illustrate the kind of problem that must be solved. A model for a laterally loaded pile is shown in Fig. 14. A pile is shown with loading at its top. Again the soil has been replaced by a set of mechanisms that conceptually define soil-response curves. Such curves give the soil resis-

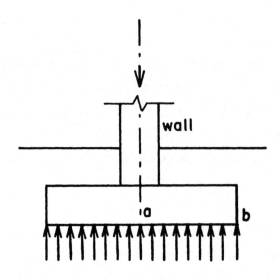

Fig. 12. Soil-structure interaction for a strip footing

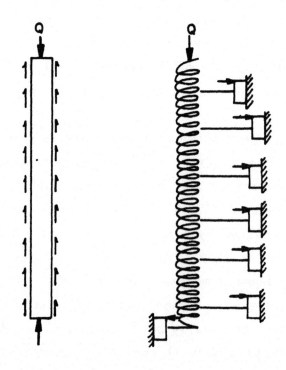

Fig. 13. Model of a pile under axial load

tance p (force per unit length along the pile) as a function of pile deflection y. The mechanisms define bilinear curves as shown in the figure, and it can be seen that the curves vary with position along the pile. Therefore, p is a nonlinear function of both y and x (refer to axes in the figure).

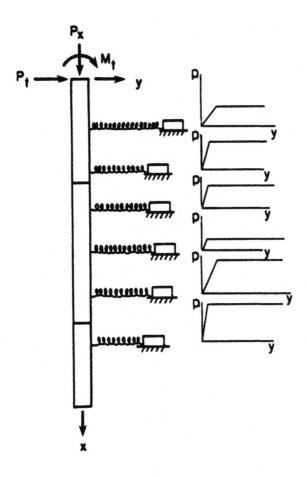

Fig. 14. Model of a pile under lateral loading showing concept of soil response curves

The problem presented in Fig. 14 is realistic except that the p-y curves for natural soils are far more complex than indicated in the figure. Later sections of the handbook will present details of methods of obtaining p-y curves and of the procedures for obtaining the deflection and bending moment along the length of the pile.

1.5 ANALYSIS OF A PILE GROUP

There are exceptions, of course, but the usual case is that piles are installed in groups. Two kinds of problems will be discussed concerning a group of piles: the loss of "efficiency" of a pile in a group of closely spaced piles, and the distribution of axial load, lateral load, and moment to a pile in a widely spaced group.

Each of these problems will be discussed in subsequent sections of this work. A considerable amount of empiricism is involved in the solution of the first of the two problems; the second one can be solved exactly if the response of a single pile to loading can be found.

1.6 GENERAL METHODS OF SOLUTION OF THE ISOLATED PILE

Five methods are available for the solution of a single pile under lateral load. These are: elastic method, curves and charts, static method (Broms method), computer method (nonlinear soil response), and nondimensional curves. The elastic method is thought to have a limited application and large numbers of curves and charts would be needed in the general case, so these two methods are not discussed herein. The other three methods are presented in appropriate detail.

The digital computer is used almost exclusively in the actual design of piles under lateral load. The method is easy to employ, a large number of variables can quickly be evaluated, and the method is relatively inexpensive.

The nondimensional method, however, has an important place in the design process. The method can be readily employed to obtain a check of the computer output. In this regard, the verification of the computer results is an essential and important aspect of the design process. There is another important function of the nondimensional procedure; it allows the engineer to gain an insight into the process involved in obtaining the response of a pile to lateral load. Such an insight is, of course, necessary to the engineer in performing designs.

The Broms static method allows an independent check of the response of a pile and also gives some insight into the solution procedure.

1.7 DESIGN PROBLEMS

The methods presented herein can be employed in the design of piles for a wide variety of applications. However, the design process involves consideration, of course, of many factors not addressed in this work and requires a suitable engineering effort.

Frequently there is some uncertainty regarding some of the parameters that enter design computations; for example, in the strength of the supporting soil. The computer method allows the influence of these parameters to be investigated with ease. If the influence of soil strength is being studied, computations can be with the upper-bound values (the absolute maximum values the shear strength could have) and the lower-bound values (the absolute minimum values). The difference in the response of the pile can be readily seen and the judgement of the designer is enhanced.

1.8 FACTOR OF SAFETY

The usual considerations concerning factor of safety enter when making a design of a pile to be subjected to lateral loading. Such things as public safety and whether a failure would cause a significant monetary loss are taken into account. However, with piles under lateral loads there are some additional considerations as discussed in the following paragraphs.

Two principal types of failure are identified herein, a soil failure and a pile failure. The soil failure is characterized by excessive deflection of the pile under lateral load and is usually associated with short penetrations. In such a case the bottom of the pile will kick toward the load as the top of the pile moves away from the load. Such action is termed "fence-posting" by some writers. The factor of safety against lateral-load failure of a short pile can be increased significantly by increasing the pile penetration. If a soil profile and a pile cross-section are selected, computations with a given lateral load show that the groundline deflection remains constant for pile penetrations beyond a critical value. As shown conceptually in Fig. 15, the groundline deflection increases rapidly as the pile penetration becomes less than the critical value. If computations for a particular design show the possibility of a soil failure, the factor of safety can be increased significantly with a small amount of additional penetration, usually with small to moderate additional cost.

A pile failure occurs when the bending moment becomes slightly greater than the bending resistance of a pile, the second type of failure asso-

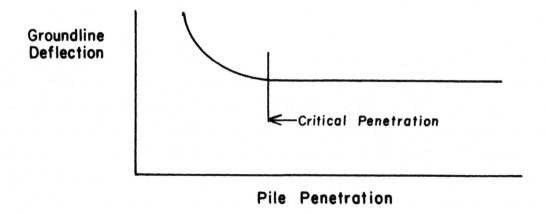

Fig. 15. Influence of pile penetration on groundline deflection

ciated with soil-structure interaction. An examination of curves of bending moment versus depth shows that in almost all cases the maximum bending moment occurs over a relatively short length of the pile. Two possibilities exist for increasing the factor of safety with little additional expense. The pile can be strengthened over the short length where the bending moment is maximum or, if the rotational restraint at the top of the pile can be controlled, an economical design can be achieved by equalizing the maximum negative and the maximum positive bending moments.

The establishment of the factor of safety for a particular design should deal with the kind of loadings to which the foundation is subjected. The AASHTO specifications deal with questions of this sort; that is, there are recommendations with regard to live load, dead load, and wind load. But in connection with lateral loadings on piles, four types of loadings are recognized: short-term static loading, repeated loading, sustained loading, and dynamic loading.

The soil response curves (p-y curves) that are recommended herein are applicable to static loading and to repeated loading. Short-term static loading rarely occurs in practice; however, that loading is included because correlations can be made with soil properties that are determined in the laboratory or the field. The curves for repeated loading are recommended because of the importance of this kind of loading. As will be shown later, repeated lateral loading of a pile can cause a considerable loss of resistance.

No recommendations are included herein for dealing with seismic loading or with sustained loading. The p-y curves for static loading will be useful in addressing the problem of the response of a pile-supported structure that is subjected to earthquake loadings. However, the dynamic analysis of a pile-supported structure involves considerations beyond the scope of this document.

With regard to sustained loading, such as will occur due to earth pressures, p-y curves for short-term loading can be used. No significant pile deflections will occur with time if the supporting soil is granular or is a stiff clay. However, consolidation will occur in soft clay and the designer must use such theory as available and judgement to estimate the additional deflection and bending moment that will occur. Obviously, it is not possible to establish a specific procedure for dealing with sustained loading because of the multitude of parameters that are involved, including the consolidation characteristics of the particular clay or silt. The selection of a factor of safety, if sustained loading occurs, must involve the use of considerable judgement.

The recommendations for p-y curves that are given later involve the use of numerous parameters describing soil properties. In all cases when subsurface soil investigations are performed, there is some uncertainty about the correct values of soil parameters to describe the soil. These uncertainties arise because of the small volume of soil that is sampled and because of unavoidable soil disturbance and other errors in soil testing. Thus, it is frequently desirable, as noted earlier, to establish upper-bound and lower-bound values for soil properties. The factor of safety for lateral-load analysis is, of course, influenced by the confidence of the engineer in the values of soil properties that are selected for design.

A companion publication (Reese, 1983) includes a number of case studies where computed and experimental results for full-scale tests are compared. A study of those comparisons will be useful to the designer to see how well the behavior at the working load agrees between experiment and computation.

1.9 INFLUENCE OF ANALYTICAL METHOD ON ENGINEERING PRACTICE

Some highway departments employ the rational method described in this report for design of piles under lateral loading. No formal survey has been made, but informal conversations with a number of highway engineers indicate that one or more of the following approximate methods are currently in use: (1) assignment of a nominal lateral load for vertical piles as recommended in building codes; (2) use of batter piles where the horizontal component of the axial load balances the horizontal forces; and (3) making computations with the Broms method. The use of the rational method should lead to improvement in designs. However, for some major projects, load tests to ascertain lateral capacities are advisable. Because of the importance of load tests, procedures for performing field load tests are presented in Chapter 7.

If the rational method employing p-y curves is used to get the response of a pile, structural engineers will have sufficient information to design the pile foundation to sustain the required loads. Combined stresses can easily be computed and reinforcement can be employed at proper positions along a pile. The reinforcement for combined stresses can consist of additional reinforcing steel, an increase in wall thickness if a pipe pile is used, or perhaps an increased diameter in some instances. If a portion of the pile extends above the groundline, the computer solution can be employed to investigate the possibility of buckling.

The approximate methods probably lead, in almost all instances, to an overdesign with regard to lateral loading. This overdesign results in increased cost. Two examples of the increased costs are given in this section.

If there is an underdesign of a pile that is subjected to lateral loading, the result will be excessive deflection or a complete collapse of the system. An example in this category is a breasting dolphin that is subjected to loads from a floating vessel. The load-deflection characteristics of breasting dolphins cannot be computed by approximate methods but can be readily computed using Computer Program COM624. The energy of the moving vessel must be balanced by the resisting energy of the dolphin and frequently the systems designed by approximate methods are so stiff that the energy absorption is overestimated. Other than the dolphin system, it is rare that a pile-supported structure is underdesigned. Rather, the usual case is overdesign with a resulting excessive cost.

Two cases are employed to illustrate the cost savings that will result from the use of the rational method. The first case is the retaining wall shown in Fig. 16. Using an approximate method, a factor of safety of one, and assuming a spacing of the pile groups of 8 ft, the horizontal load that must be taken by Pile B is 20.7 kips, resulting in an axial load of 82.8 kips. The equations of static equilibrium show Pile A is to be lightly loaded. While it is known that both Piles A and B will be subjected to bending, the equations of statics do not result in bending moment diagrams; therefore, bending resistance must be built into Piles A and B in order to ensure stability.

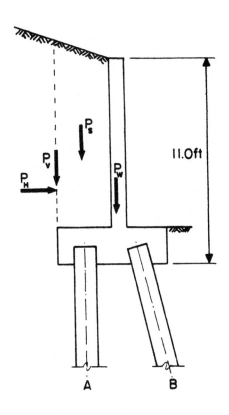

Fig. 16. Retaining wall for example computation

Using the rational method and making the same assumptions about the factor of safety and the pile spacing, the axial load on Pile B is computed to be 38.9 kips, and the lateral load is computed to be 5.8 kips. The lateral load on Pile A is about the same and the maximum bending moment on each of the piles is computed to be about 145 in-kips. A curve showing moment versus depth can be computed if it is desirable to vary the reinforcement of the piles with depth. The opportunities for savings are obvious. A thorough study of the system with the rational method would achieve an optimum design. Furthermore, details such as the way the pile is fastened to the concrete mat can be designed for maximum benefit. It appears that the rational method would cut the number of piles by about 50% over what would be required when employing the approximate method.

The second example concerns the foundations for an overhead sign support. The foundation can consist of two piles (drilled shafts) as shown in Fig. 17a or by one drilled shaft as shown in Fig. 17b. In the first case, the overturning moment is taken by tension in the windward pile and by compression in the leeward pile. In an example that has been worked out (Lowe and Reese, 1982), the two-pile system would cost approximately $5700 for four 30-inch-diameter drilled shafts with a penetration of 19 ft while the single-shaft system would cost about $3750 for two 30-inch-diameter drilled shafts with a penetration of 25 ft. Thus, the savings of the single-shaft system, designed by the rational method, would be significant if employed throughout a highway system. It is worthy of note, however, that overhead-sign contractors could place the shorter shafts while foundation contractors might be required to place the longer shafts.

1.10 ORGANIZATION OF HANDBOOK

The material presented herein is intended to be a convenient and useful guide to the design of single piles and pile groups under lateral loading. A chapter on soil response is followed by three chapters on methods of analysis of a single pile. The computer method is emphasized. Chapter 6 concerns the design of pile groups.

While the material presented herein is a sufficient guide for design, a review of the references cited will be useful. Futhermore, design engineers are encouraged to stay abreast with the technical literature because additional data from lateral-load tests are becoming available from time to time.

1.11 DESIGN ORGANIZATION

A comment is desirable about the design team that is responsible for the design of piles under lateral loading. One possibility is that the geotechnical engineer could provide soil-response curves to the structural engineer who would do the computations of bending moment and deflection. A better arrangement is that the geotechnical engineer and

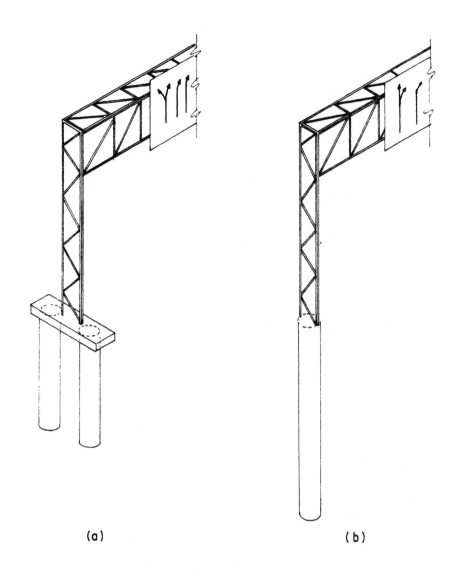

Fig. 17. Sign supports for example computations

the structural engineer work together during the design process. The influence of altering pile penetration, diameter, and stiffness and loading conditions could be considered jointly. A more favorable design would then be possible.

CHAPTER 2. SOIL RESPONSE

2.1 NATURE OF SOIL RESPONSE

The manner in which the soil responds to the lateral deflection of a pile can be examined by considering the pipe pile shown in Fig. 18. Two slices of soil are indicated; the element A is near the ground surface and the element B is several diameters below the ground surface. Consideration will be given here to the manner in which those two elements of soil react as the pile deflects under an applied lateral load.

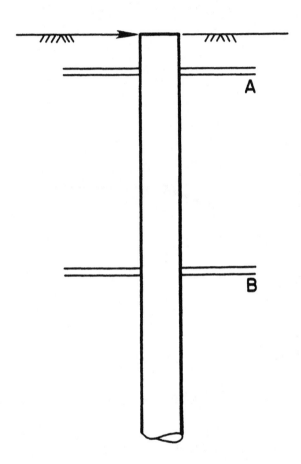

Fig. 18. Pipe pile and soil elements

In each case the desired information is the force per unit length, similar to the loading used in solving a beam problem. As the pile moves in response to the loading, the force on the back side of the pile will decrease and that on the front side will increase. The result is the soil response. As shown later, experiments with instrumented piles show that the shape of the soil response curves (p-y curves) can be complex; however, some useful concepts can be set forth from elementary considerations.

Figure 19 shows a p-y curve that is conceptual in nature. The curve is plotted in the first quadrant for convenience and only one branch is shown. The curve properly belongs in the second and fourth quadrants because the soil response acts in opposition to the deflection. The branch of the p-y curve O-a is representative of the elastic action of the soil; the deflection at point a may be small. The branch a-b is the transition portion of the curve. At point b the ultimate soil resistance is reached. The following paragraphs will deal with the ultimate soil resistance.

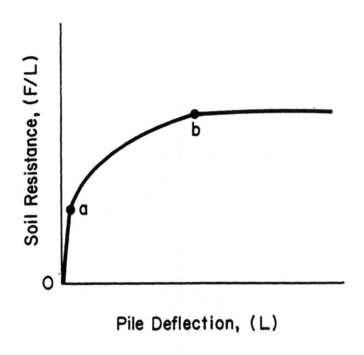

Fig. 19. Conceptual p-y curve

With regard to the ultimate resistance at element A in Fig. 18, Fig. 20 shows a wedge of soil that is moved up and away from a pile. The magnitude of the force F_p can be computed by using the weight of the wedge and by assuming that the soil on the sliding surfaces is at failure. The above approach does not take into account the stress on the back side of the pile that can easily be taken into account but, in any case, is small.

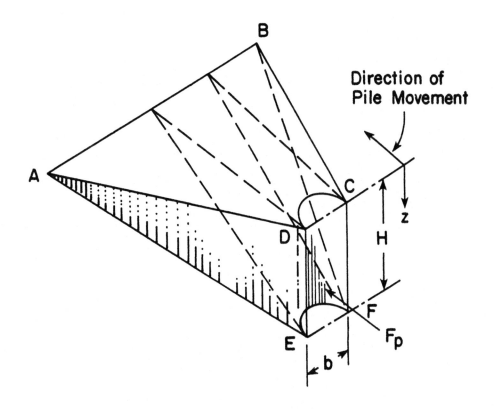

Fig. 20. Wedge-type failure of surface soil

Numerical evaluation of the ultimate soil resistance per unit length of the pile p_u can be made by use of the wedge shown in Fig. 20. The ground surface is represented by the plane ABCD and the soil in contact with the pile is represented by the surface CDEF. If the pile is moved in the direction indicated, the wedge of soil will move up and out and failure of the soil in shear will result on the planes ADE, BCF, and AEFB. The horizontal force F_p against the pile can be computed by summing the horizontal components of the forces on the sliding surfaces, taking into account the gravity force on the wedge of soil. For a given value of H, it is assumed that the value of the horizontal force on the pile is F_{p1}. If a second computation is made with the depth of the wedge increased by ΔH, the horizontal force will be F_{p2}. The value of p_u for the depth z where z is equal approximately to $(2H + \Delta H)/2$ can be computed: $(p_u)_z = (F_{p2} - F_{p1})/\Delta H$. Another approach is to write an expression for F_p in terms of z and to differentiate the expression to find p_u as a function of z.

At the ground surface, the value of p_u for sand must be zero because the weight of the wedge is zero and the forces on the sliding surfaces will be zero. At the ground surface for clay, on the other hand, the value of p_u will be larger than zero because the cohesion of the clay, which is independent of the overburden stress, will generate a horizontal force.

A plan view of a pile at several diameters below the ground surface, corresponding to the element at B in Fig. 18, is shown in Fig. 21. The potential failure surfaces that are shown are indicative of plane-strain failure; while the ultimate resistance p_u cannot be determined precisely, elementary concepts can be used to develop approximate expressions.

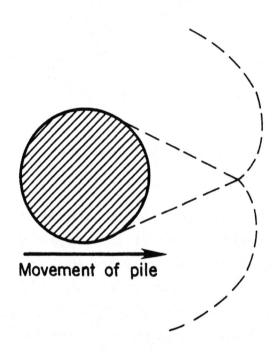

Fig. 21. Potential failure surfaces generated by pile at several diameters below ground surface

2.2 EFFECTS OF LOADING AND PRESENCE OF WATER

As will be shown in detail in the next sections, the soil response can be affected by the way the load is applied to a pile. As noted earlier, recommendations are given herein for the cases where the load is short-term (static) or is repeated (cyclic). The latter case is frequently encountered in design. Loadings that are sustained or dynamic (due to machinery or a seismic event) are special cases; the methods of dealing with these types of loading are not well developed and are not addressed herein.

The presence of water will affect the unit weight of the soil and will perhaps affect other properties to some extent; however, water above the ground surface has a pronounced effect on the response of clay soils, particularily stiff clay. Cyclic loading has two types of deleterious effects on clays; there is likely to be strain softening due to repeated deformations and scour at the pile-soil interface. This latter effect can be the most serious. If the deflection of the pile is greater than that at point a in Fig. 19 or certainly if the deflection is greater than that at point b, a space will open as the load is released. The space will fill with water and the water will be pushed upward, or through cracks in the clay, with the next cycle of loading. The velocity of the water can be such that considerable quantities of soil are washed to the ground surface, causing a significant loss in soil resistance.

The cyclic loading of sands also causes a reduced resistance in sands but the reduction is much less severe than experienced by clays.

2.3 METHOD OF TREATING A LAYERED SOIL PROFILE

As will be noted in the sections that follow, no specific recommendations are given for p-y curves for layered soils. However, some research (Hargrove, 1981) has indicated that each individual layer can be treated as if the entire stratum consists of soil with the characteristics of that layer. For example, if a layer of clay underlies a surface layer of sand, the p-y curves of the clay will be computed as if the clay exists to the ground surface.

There is an adjustment that needs to be made, however, for the case described above in computing the p-y curves. It will be necessary to give the sand ficticious values of cohesion c with the ficticious values of c computed by multiplying the vertical effective stress by the tangent of the angle of internal friction. The ficticious values of c, being zero at the ground surface and increasing with depth, will not change the response of the upper sand stratum because the computation of the p-y curves for sand depends only on the strength parameter ϕ, the angle of internal friction. The adjustment will allow the p_u to be computed correctly using the wedge model in Fig. 20.

27

The remaining sections of this chapter present detailed methods for obtaining p-y curves. Recommendations are given for clay and sand, for static and cyclic loading, and cases where the water table is above or below the ground surface. As will be seen, the soil properties that are needed for clay refer to undrained shear strength; there are no provisions for dealing with a c-ϕ soil.

The remaining portions of this chapter follow closely the presentation of p-y curves as presented in the manual, "Behavior of Piles and Pile Groups under Lateral Load," prepared for Federal Highway Administration, Office of Research, Washington, D.C., 1983.

2.4 p-y CURVES FOR SOFT CLAY

As noted earlier, there is a significant influence of the presence of water above the ground surface. If soft clay exists at the ground surface, it is obvious that water must be present at or above the ground surface or the clay would have become desiccated and stronger. If soft clay does not exist at the ground surface but exists at some distance below the ground surface, the deleterious effect of water moving in and out of a gap at the interface of the pile and soil will not occur; therefore, the p-y curves for clay above the ground surface should probably be used (Welch and Reese, 1972). The p-y curves presented here for soft clay were developed with water above the ground surface and experienced the deteriorating effects discussed earlier.

Field Experiments

Field experiments using full-sized, instrumented piles provide data from which p-y curves from static and from cyclic loading can be generated. Such experimental curves are correlated with available theory and the correlations provide the basis for recommendations for procedures for developing p-y curves. Therefore, field experiments with instrumented piles are essential to the recommendations for p-y curves and the relevant reports on field tests will be referenced.

Matlock (1970) performed lateral load tests employing a steel-pipe pile that was 12.75 inches in diameter and 42 ft long. It was driven into clays near Lake Austin that had a shear strength of about 800 lb/sq ft. The pile was recovered, taken to Sabine Pass, Texas, and driven into clay with a shear strength that averaged about 300 lb/sq ft in the significant upper zone. The studies carried out by Matlock led to the recommendations shown in the following paragraphs.

Recommendations for Computing p-y Curves

The following procedure is for <u>short-term static loading</u> and is illustrated by Fig. 22a.

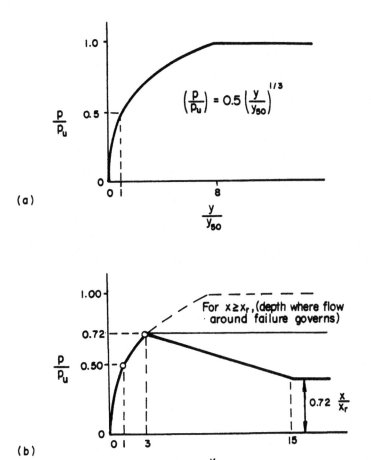

Fig. 22. Characteristic shapes of the p-y curves
for soft clay below the water table
(a) for static loading
(b) for cyclic loading
(from Matlock, 1970)

1. Obtain the best possible estimate of the variation with depth of undrained shear strength c and submerged unit weight γ'. Also obtain the values of ε_{50}, the strain corresponding to one-half the maximum principal-stress differ-

29

ence. If no stress-strain curves are available, typical values of ε_{50} are given in the following table.

TABLE 1. REPRESENTATIVE VALUES OF ε_{50}

Consistency of Clay	ε_{50}
Soft	0.020
Medium	0.010
Stiff	0.005

(Also see Tables 3 and 4)

2. Compute the ultimate soil resistance per unit length of pile, using the smaller of the values given by equations below.

$$p_u = \left[3 + \frac{\gamma'}{c} x + \frac{J}{b} x \right] cb \tag{1}$$

$$p_u = 9 cb \tag{2}$$

where

$\quad x$ = depth from ground surface to p-y curve

$\quad \gamma'$ = average effective unit weight from ground surface to depth x

$\quad c$ = shear strength at depth x

$\quad b$ = width of pile

$\quad J$ = empirical parameter.

Matlock (1970) states that the value of J was determined experimentally to be 0.5 for a soft clay and about 0.25 for a medium clay. A value of 0.5 is frequently used for J. The value of p_u is computed at each depth where a p-y curve is desired, based on shear strength at that depth. The computer obtains values of y and the corresponding p-values at close spacings; if hand computations are being done, p-y curves should be computed at depths to reflect the soil profile. If the soil is homogeneous, the p-y curves should be obtained at close spacings near the ground surface where the pile deflection is greater.

3. Compute the deflection, y_{50}, at one-half the ultimate soil resistance from the following equation:

$$y_{50} = 2.5 \varepsilon_{50} b. \tag{3}$$

4. Points describing the p-y curve are now computed from the following relationship.

$$\frac{p}{p_u} = 0.5 \left(\frac{y}{y_{50}}\right)^{\frac{1}{3}} \tag{4}$$

The value of p remains constant beyond $y = 8y_{50}$.

The following procedure is for <u>cyclic loading</u> and is illustrated in Fig. 22b.

1. Construct the p-y curve in the same manner as for short-term static loading for values of p less than $0.72p_u$.
2. Solve Eqs. 1 and 2 simultaneously to find the depth, x_r, where the transition occurs from the wedge-type failure to a flow-around failure. If the unit weight and shear strength are constant in the upper zone, then

$$x_r = \frac{6 cb}{(\gamma'b + Jc)}. \tag{5}$$

If the unit weight and shear strength vary with depth, the value of x_r should be computed with the soil properties at the depth where the p-y curve is desired.

3. If the depth to the p-y curve is greater than or equal to x_r, then p is equal to $0.72p_u$ for all values of y greater than $3y_{50}$.
4. If the depth to the p-y curve is less than x_r, then the value of p decreases from $0.72p_u$ at $y = 3y_{50}$ to the value given by the following expression at $y = 15y_{50}$.

$$p = 0.72 \, p_u \left(\frac{x}{x_r}\right) \tag{6}$$

The value of p remains constant beyond $y = 15y_{50}$.

Recommended Soil Tests

For determining the various shear strengths of the soil required in the p-y construction, Matlock (1970) recommended the following tests in order of preference:

1. in-situ vane-shear tests with parallel sampling for soil identification,
2. unconsolidated-undrained triaxial compression tests having a confining stress equal to the overburden pressure with c being defined as half the total maximum principal stress difference,
3. miniature vane tests of samples in tubes, and
4. unconfined compression tests.

Matlock's recommendations are related to the determination of the shear strength of the clay. So if field vane tests are performed, for example, other tests must be performed to obtain ε_{50} and the unit weight of the soil.

In the test program, emphasis should be placed on determining soil properties at and near the ground surface. In the general case, the farther the soil from the ground surface the less is its influence on pile response. For a soil profile that is relatively homogeneous, studies have shown that the soil below ten pile diameters has only a small influence on pile deflection and bending moment.

Example Curves

An example set of p-y curves was computed for soft clay for a pile with a diameter of 48 in. The soil profile that was used is shown in Fig. 23. The submerged unit weight was assumed to be 20 lb/cu ft at the mudline and 40 lb/cu ft at a depth of 80 ft and to vary linearly. In the absence of a stress-strain curve for the soil, ε_{50} was taken as 0.01 for the full depth of the soil profile. The loading was assumed to be cyclic.

The p-y curves were computed for the following depths below the mudline: 0, 1, 2, 4, 8, 12, 20, 40, and 60 ft. The plotted curves are shown in Fig. 24. (Curves for 0 and 1 ft too close to axis to be shown.)

The step-by-step procedure for computing the cyclic p-y curve at a depth of 20 ft is as follows:

1. Obtain the best possible estimate of the following soil parameters:

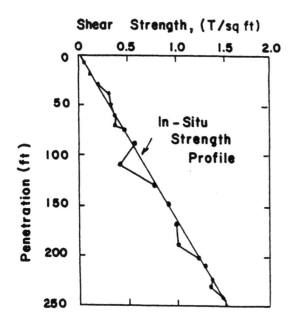

Fig. 23. Soil profile used for example p-y curve for soft clay

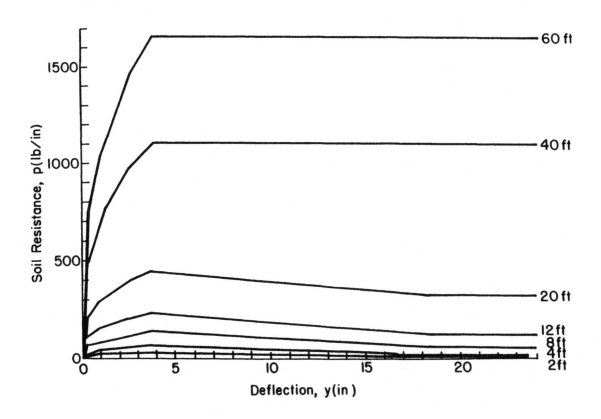

Fig. 24. Example p-y curves for soft clay below water table, Matlock criteria, cyclic loading

at depth 20 ft

$$c = (20)\left(\frac{1.5}{250}\right) = 0.12 \text{ T/sq ft} = 240 \text{ lb/sq ft}$$

$$\gamma' = 20 + (40 - 20)\left(\frac{20}{80}\right) = 25 \text{ lb/cu ft}$$

$$\varepsilon_{50} = 0.01.$$

2. Compute the ultimate soil resistance per unit length of pile:

   From Eq. 1, $\quad p_u = \left[3 + \frac{\gamma'}{c} x + \frac{J}{b} x\right] cb$

   (Pile width b = 4 ft = 48 in, J = 0.5)

   $$p_u = \left[3 + \left(\frac{25}{240}\right)(20) + \left(\frac{0.5}{4}\right)(20)\right](240)(4)$$

   $\quad\quad = 7280 \text{ lb/ft} = 607 \text{ lb/in.}$

   From Eq. 2, $\quad p_u = 9cb$

   $\quad\quad\quad = (9)(240)(4)$

   $\quad\quad\quad = 8640 \text{ lb/ft} = 720 \text{ lb/in.}$

   Select the smaller value from the above computation

   $p_u = 607 \text{ lb/in.}$

3. Compute the deflection, y_{50}:

   From Eq. 3, $\quad y_{50} = 2.5 \, \varepsilon_{50} \, b$

   $\quad\quad\quad\quad = (2.5)(0.01)(48) = 1.2 \text{ in.}$

4. Compute the transition depth x_r:

   From Eq. 5,

   $$x_r = \frac{6cb}{\gamma' b + Jc} = \frac{(6)(240)(4)}{(25)(4) + (0.5)(240)} = 26.18 \text{ ft}$$

5. Because the depth to the p-y curve (x = 20 ft) is less than x_r (x_r = 26.18 ft):

   From Eq. 4, $\quad p = (p_u)(0.5)\left(\frac{y}{y_{50}}\right)^{\frac{1}{3}} \quad\quad 0 < y < 3y_{50}$

   $\quad\quad\quad\quad\quad p = 0.72 \, p_u \quad\quad\quad\quad\quad\quad y = 3y_{50}$

   From Eq. 6, $\quad p = 0.72 \, p_u (0.5)\left(\frac{x}{x_r}\right) \quad\quad y = 15y_{50}$

   $3y_{50} = (3)(1.2) = 3.6 \text{ in}$

   $15y_{50} = (15)(1.2) = 18 \text{ in}$

   $0.72 \, p_u = (0.72)(607) = 437 \text{ lb/in.}$

6. Points describing the p-y curve are now computed from the above relationship (shown in step 5)

y (in)	p (lb/in)
0.01	$p = (607)(0.5)\left(\frac{0.01}{1.2}\right)^{\frac{1}{3}} = 61.5$
0.60	$p = (607)(0.5)\left(\frac{0.60}{1.20}\right)^{\frac{1}{3}} = 240.9$
1.20	$p = (607)(0.5)\left(\frac{1.20}{1.20}\right)^{\frac{1}{3}} = 303.5$
2.40	$p = (607)(0.5)\left(\frac{2.40}{1.20}\right)^{\frac{1}{3}} = 382.4$
3.60	$p = 0.72\, p_u = 437.0$
18.00	$p = 0.72\, p_u \frac{x}{x_r} = 0.72(607)\left(\frac{20}{26.18}\right) = 334$

2.5 p-y CURVES FOR STIFF CLAY BELOW THE WATER TABLE

Field Experiments

Reese, Cox, and Koop (1975) performed lateral load tests employing steel-pipe piles that were 24 inches in diameter and 50 ft long. The piles were driven into stiff clay at a site near Manor, Texas. The clay had an undrained shear strength ranging from about 1 T/sq ft at the ground surface to about 3 T/sq ft at a depth of 12 ft. The studies that were carried out led to the recommendations shown in the following paragraphs.

Recommendations for Computing p-y Curves

The following procedure is for <u>short-term static loading</u> and is illustrated by Fig. 25. The empirical parameters, A_s and A_c, shown in Fig. 26 and k_s and k_c shown in Table 2 were determined from the results of the experiments.

1. Obtain values for undrained soil shear strength c, soil submerged unit weight γ', and pile diameter b.
2. Compute the average undrained soil shear strength c_a over the depth x.
3. Compute the ultimate soil resistance per unit length of pile using the smaller of the values given by the equation

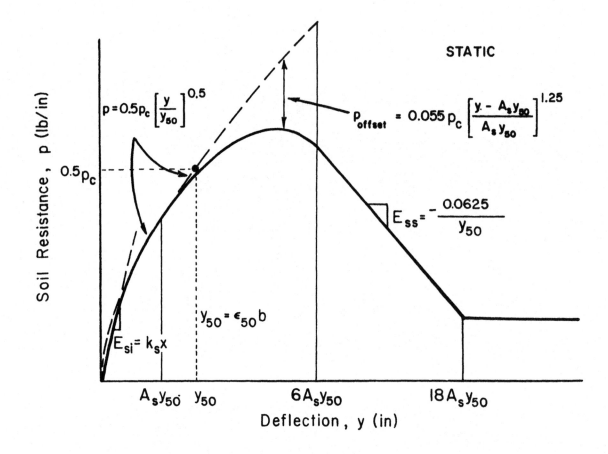

Fig. 25. Characteristic shape of p-y curve for static loading in stiff clay below the water table (after Reese, Cox, Koop, 1975)

below

$$p_{ct} = 2c_a b + \gamma' bx + 2.83 c_a x, \quad (7)$$
$$p_{cd} = 11\, cb. \quad (8)$$

4. Choose the appropriate value of A_s from Fig. 26 for the particular nondimensional depth.

5. Establish the initial straight-line portion of the p-y curve:

$$p = (kx)y. \quad (9)$$

Use the appropriate value of k_s or k_c from Table 2 for k.

6. Compute the following:

$$y_{50} = \varepsilon_{50} b. \quad (10)$$

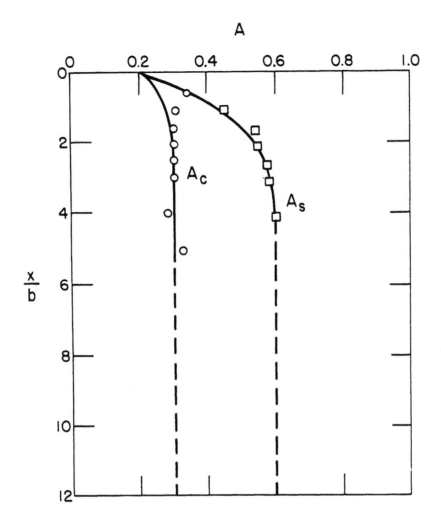

Fig. 26. Values of constants A_s and A_c

TABLE 2. REPRESENTATIVE VALUES OF k FOR STIFF CLAYS

	Average Undrained Shear Strength* T/sq ft		
	0.5-1	1-2	2-4
k_s (Static) lb/cu in	500	1000	2000
k_c (Cyclic) lb/cu in	200	400	800

*The average shear strength should be computed to a depth of 5 pile diameters. It should be defined as half the total maximum principal stress difference in an unconsolidated undrained triaxial test.

(Also see Table 6)

Use an appropriate value of ε_{50} from results of laboratory tests or, in the absence of laboratory tests, from Table 3.

TABLE 3. REPRESENTATIVE VALUES OF ε_{50} FOR STIFF CLAYS

	Average Undrained Shear strength T/sq ft		
	0.5-1	1-2	2-4
ε_{50} (in/in)	0.007	0.005	0.004

(Also see Tables 1 and 4)

7. Establish the first parabolic portion of the p-y curve, using the following equation and obtaining p_c from Eqs. 7 or 8.

$$p = 0.5 p_c \left(\frac{y}{y_{50}}\right)^{0.5} \quad (11)$$

Equation 11 should define the portion of the p-y curve from the point of the intersection with Eq. 9 to a point where y is equal to $A_s y_{50}$ (see note in step 10).

8. Establish the second parabolic portion of the p-y curve,

$$p = 0.5 p_c \left(\frac{y}{y_{50}}\right)^{0.5} - 0.055 p_c \left(\frac{y - A_s y_{50}}{A_s y_{50}}\right)^{1.25} \quad (12)$$

Equation 12 should define the portion of the p-y curve from the point where y is equal to $A_s y_{50}$ to a point where y is equal to $6 A_s y_{50}$ (see note in step 10).

9. Establish the next straight-line portion of the p-y curve,

$$p = 0.5 p_c (6 A_s)^{0.5} - 0.411 p_c - \left(\frac{0.0625}{y_{50}}\right) p_c (y - 6 A_s y_{50}). \quad (13)$$

Equation 13 should define the portion of the p-y curve from the point where y is equal to $6A_s y_{50}$ to a point where y is equal to $18A_s y_{50}$ (see note in step 10).

10. Establish the final straight-line portion of the p-y curve,

$$p = 0.5 p_c (6A_s)^{0.5} - 0.411 p_c - 0.75 p_c A_s \quad (14)$$

or

$$p_c = p_c (1.225 \sqrt{A_s} - 0.75 A_s - 0.411). \quad (15)$$

Equation 15 should define the portion of the p-y curve from the point where y is equal to $18A_s y_{50}$ and for all larger values of y (see following note).

Note: The step-by-step procedure is outlined, and Fig. 25 is drawn, as if there is an intersection between Eqs. 9 and 11. However, there may be no intersection of Eq. 9 with any of the other equations defining the p-y curve. If there is no intersection, the equation should be employed that gives the smallest value of p for any value of y.

The following procedure is for <u>cyclic loading</u> and is illustrated in Fig. 27.

1. Steps 1, 2, 3, 5, and 6 are the same as for the static case.
4. Choose the appropriate value of A_c from Fig. 26 for the particular nondimensional depth.

Compute the following:

$$y_p = 4.1 A_s y_{50}. \quad (16)$$

7. Establish the parabolic portion of the p-y curve,

$$p = A_c p_c \left[1 - \left| \frac{y - 0.45 y_p}{0.45 y_p} \right|^{2.5} \right] \quad (17)$$

Equation 17 should define the portion of the p-y curve from the point of the intersection with Eq. 9 to where y is equal to $0.6 y_p$ (see note in step 9).

8. Establish the next straight-line portion of the p-y curve,

$$p = 0.936 A_c p_c - \frac{0.085}{y_{50}} p_c (y - 0.6 y_p). \quad (18)$$

Equation 18 should define the portion of the p-y curve from the point where y is equal to $0.6 y_p$ to the point where y is equal to $1.8 y_p$ (see note in step 9).

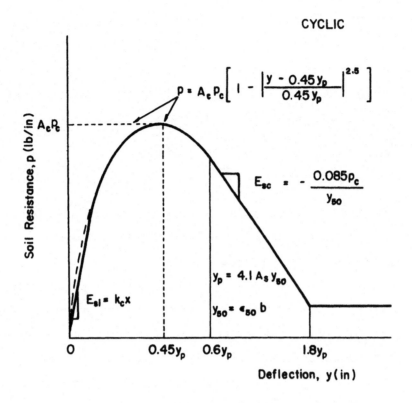

Fig. 27. Characteristic shape of p-y curve for cyclic loading in stiff clay below water table (after Reese, Cox, Koop, 1975)

9. Establish the final straight-line portion of the p-y curve,

$$p = 0.936 \, A_c p_c - \frac{0.102}{y_{50}} p_c y_p. \tag{19}$$

Equation 19 should define the portion of the p-y curve from the point where y is equal to $1.8y_p$ and for all larger values of y (see following note).

Note: The step-by-step procedure is outlined, and Fig. 27 is drawn, as if there is an intersection between Eqs. 9 and 17. However, there may be no intersection of those two equations and there may be no intersection of Eq. 9 with any of the other equations defining the p-y curve. If there is no intersection, the equation should be employed that gives the smallest value of p for any value of y.

Recommended Soil Tests

Triaxial compression tests of the unconsolidated-undrained type with confining pressures conforming to the in-situ total overburden pressures are recommended for determining the shear strength of the soil. The value of ε_{50} should be taken as the strain during the test corresponding to the stress equal to half the maximum total-principal-stress difference. The shear strength, c, should be interpreted as one-half of the maximum total-stress difference. Values obtained from the triaxial tests might be somewhat conservative but would represent more realistic strength values than other tests. The unit weight of the soil must be determined.

Example Curves

An example set of p-y curves was computed for stiff clay for a pile with a diameter of 48 in. The soil strength profile that was used is shown in Fig. 28. The submerged unit weight of the soil was assumed to be 50 lb/cu ft for the entire depth. In the absence of a stress-strain curve, ε_{50} was taken as 0.005 for the full depth of the soil profile. The slope of the initial portion of the p-y curve was established by assuming a value of k of 463 lb/cu in. The loading was assumed to be cyclic.

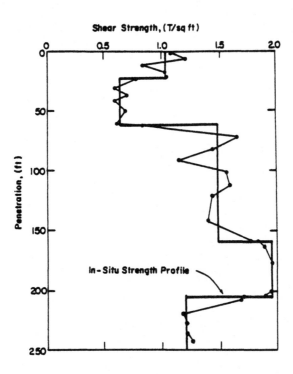

Fig. 28. Soil profile used for example p-y curves for stiff clay

The p-y curves were computed for the following depths below the mudline: 0, 2, 4, 8, 12, 20, 40, and 60 ft. The plotted curves are shown in Fig. 29.

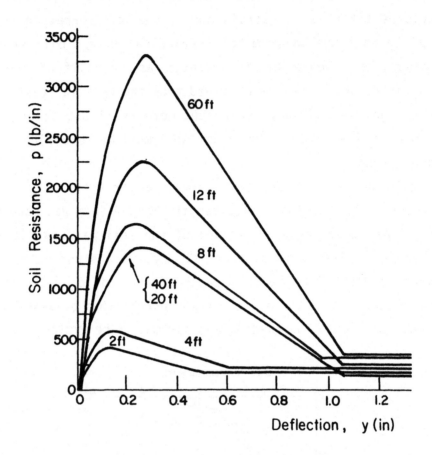

Fig. 29. Example p-y curves for stiff clay below the water table, Reese criteria, cyclic loading

The step-by-step solutions for computing the cyclic p-y curve at a depth of 12 ft are shown as follows:

1. Obtain the best possible estimate of the following soil parameters:

    at depth 12 ft

    c = 1.05 T/sq ft = 2100 lb/sq ft = 14.58 lb/sq in

    γ' = 50 lb/cu ft

    ε_{50} = 0.005

    k = 463 lb/cu in. (See Table 2)

2. Compute the average undrained soil shear strength c_a over the depth x:

   $c = 2100$ lb/sq ft $x = 0$
   $c = 2100$ lb/sq ft $x = 12$ ft
   $c_a = \frac{2100 + 2100}{2} = 2100$ lb/sq ft

3. Compute the ultimate soil resistance per unit length of pile:

   From Eq. 7, $p_{ct} = 2c_a b + \gamma'bx + 2.83\ c_a x$
   $= (2)(2100)(4) + (50)(4)(12) + (2.83)(2100)(12)$
   $= 90516$ lb/ft $= 7543$ lb/in.

   From Eq. 8, $p_{cd} = 11\ cb$
   $= (11)(2100)(4) = 92400$ lb/ft
   $= 7700$ lb/in.

   Select $p_c = p_{ct} = 7543$ lb/in.

4. Establish the initial straight-line portion of the p-y curve:

   From Eq. 9, $p = (kx)y = (463)(144)y$
   $= 66672y$ (up to y_1)

5. Compute y_{50}:

   From Eq. 10, $y_{50} = \varepsilon_{50} b = (0.005)(48) = 0.24$ in.

6. Compute y_p:

   Select A_s as 0.59 from Fig. 26.

   From Eq. 16, $y_p = 4.1 A_s y_{50}$
   $= (4.1)(0.59)(0.24) = 0.581$ in.

7. Establish the parabolic portion of the p-y curve:

   From Eq. 17, $p = A_c p_c \left[1 - \left|\frac{y - 0.45 y_p}{0.45 y_p}\right|^{2.5}\right]$

   $y_1 < y \leq 0.6 y_p$

y_1 is the switch point from straight-line portion to the parabolic portion and can be solved from Eq. 9 and Eq. 17.

Since there is no common solution for y_1 between Eq. 9 and Eq. 17 in this case, we should take the smallest value of p for any value of y from Eq. 9 and 17 for constructing the initial portion of p-y curve.

From Eq. 9, $p = (kx)y$

For $y = 0.05$ in, $p = (463)(144)(0.05)$

$p = 3334$ lb/in

From Eq. 17, $p = A_c p_c \left[1 - \left| \dfrac{y - 0.45 y_p}{0.45 y_p} \right|^{2.5} \right]$

$y_1 < y < 0.6 y_p$

For $y = 0.05$ in, $p = (0.3)(7543)\left[1 - \left| \dfrac{0.5 - (0.45)(0.581)}{(0.45)(0.581)} \right|^{2.5} \right]$

$p = 932$ lb/in.

y_1 (in)	p (from Eq. 9) (lb/in)	p (from Eq. 17) (lb/in)	Selected p (lb/in)
0.050	3334	932	932
0.100	6668	1585	1585
0.200	13334	2202	2202
0.300	20002	2244	2244
0.348	23202	2120	2120

8. Establish the next straight-line portion of the p-y curve:

From Eq. 18, $p = 0.936 A_c p_c - \dfrac{0.085}{y_{50}} p_c (y - 0.6 y_p)$

$0.6 y_p < y < 1.8 y_p$

For $y = 1.046$ in, $p = (0.936)(0.3)(7543)$

$- \dfrac{0.085}{0.240}(7543)[1.046 - (0.6)(0.581)]$

$p = 255$ lb/in.

y (in)	p (lb/in)	
0.35	2114	
0.40	1981	$0.6 y_p = 0.349$
0.60	1446	$1.8 y_p = 1.046$
1.046	255	

9. Establish the final straight-line portion of the p-y curve:

From Eq. 19, $p = 0.936 A_c p_c - \dfrac{0.102}{y_{50}} p_c y_p$

$y \geq 1.8 y_p$

$p = (0.936)(0.3)(7543)$
$\quad - (\dfrac{0.102}{0.24})(7543)(0.581)\quad y \geq 1.046$
$= 256 \text{ lb/in.}$

2.6 p-y CURVES FOR STIFF CLAY ABOVE THE WATER TABLE

Field Experiments

A lateral load test was performed at a site in Houston on a drilled shaft, 36 inches in diameter. A 10-in diameter pipe, instrumented at intervals along its length with electrical-resistance-strain gauges, was positioned along the axis of the shaft before concrete was placed. The embedded length of the shaft was 42 ft. The average undrained shear strength of the clay in the upper 20 ft was approximately 2,200 lb/sq ft. The experiments and their intepretation are discussed in detail by Welch and Reese (1972) and Reese and Welch (1975). The results of the experiments were used to develop recommendations for p-y curves that are shown in the following paragraphs.

Recommendations for Computing p-y Curves

The following procedure is for <u>short-term static loading</u> and is illustrated in Fig. 30a.

1. Obtain values for undrained shear strength c, soil unit weight γ, and pile diameter b. Also obtain the values of ε_{50} from stress-strain curves. If no stress-strain curves are available, use a value from ε_{50} of 0.010 or 0.005 as given in Table 1, the larger value being more conservative.

2. Compute the ultimate soil resistance per unit length of shaft, p_u, using the smaller of the values given by Eqs. 1 and 2. (In the use of Eq. 1 the shear strength is taken as the average from the ground surface to the depth being considered and J is taken as 0.5. The unit weight of the soil should reflect the position of the water table.)

3. Compute the deflection, y_{50}, at one-half the ultimate soil resistance from Eq. 3.

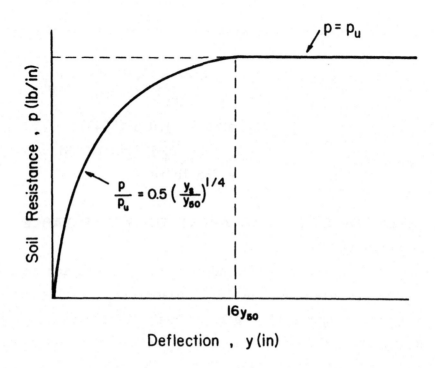

Fig. 30a. Characteristic shape of p-y curve for static loading in stiff clay above water table

4. Points describing the p-y curve may be computed from the relationship below.

$$\frac{p}{p_u} = 0.5 \left(\frac{y}{y_{50}}\right)^{\frac{1}{4}} \qquad (20)$$

5. Beyond $y = 16y_{50}$, p is equal to p_u for all values of y.

The following procedure is for <u>cyclic loading</u> and is illustrated in Fig. 30b.

1. Determine the p-y curve for short-term static loading by the procedure previously given.
2. Determine the number of times the design lateral load will be applied to the pile.
3. For several values of p/p_u obtain the value of C, the parameter describing the effect of repeated loading on deformation, from a relationship developed by laboratory tests,

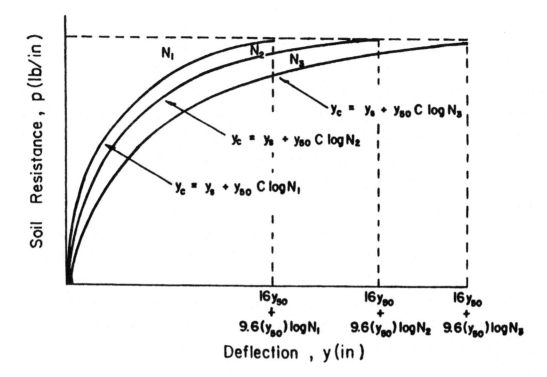

Fig. 30b. Characteristic shape of p-y curve for cyclic loading in stiff clay above water table

(Welch and Reese, 1972), or in the absence of tests, from the following equation.

$$C = 9.6 \left(\frac{p}{p_u}\right)^4 \tag{21}$$

4. At the value of p corresponding to the values of p/p_u selected in step 3, compute new values of y for cyclic loading from the following equation.

$$y_c = y_s + y_{50} \cdot C \cdot \log N \tag{22}$$

where

y_c = deflection under N-cycles of load,

y_s = deflection under short-term static load,

y_{50} = deflection under short-term static load at one-half the ultimate resistance, and

N = number of cycles of load application

5. The p-y curve defines the soil response after N-cycles of load.

Recommended Soil Tests

Triaxial compression tests of the unconsolidated-undrained type with confining stresses equal to the overburden pressures at the elevations from which the samples were taken are recommended to determine the shear strength. The value of ε_{50} should be taken as the strain during the test corresponding to the stress equal to half the maximum total principal stress difference. The undrained shear strength, c, should be defined as one-half the maximum total-principal-stress difference. The unit weight of the soil must also be determined.

Example Curves

An example set of p-y curves was computed for stiff clay above the water table for a pile with a diameter of 48 in. The soil profile that was used is shown in Fig. 28. The unit weight of the soil was assumed to be 112 lb/cu ft for the entire depth. In the absence of a stress-strain curve, ε_{50} was taken as 0.005. Equation 21 was used to compute values for the parameter C and it was assumed that there is to be 100 cycles of load application.

The p-y curves were computed for the following depths below the groundline: 0, 1, 2, 4, 8, 12, 20, 40, and 60 ft. The plotted curves are shown in Fig. 31.

The step-by-step solutions for computing the cyclic p-y curve at a depth of 40 ft are shown as follows:

1. Obtain the best possible estimate of soil parameters at depth 40 ft:

    b = 48 in
    c = 0.65 T/sq ft = 1300 lb/sq ft
    γ = 112 lb/cu ft
    ε_{50} = 0.005
    N = 100 cycles
    c_{avg} = 0.85 T/sq ft = 1700 lb/sq ft = 11.80 lb/sq in
    c@40' = 0.65 T/sq ft = 1300 lb/sq ft = 9.03 lb/sq in.

2. Compute the ultimate soil resistance per unit length of pile:

    From Eq. 1,
    $$p_u = \left[3 + \frac{\gamma'}{c_{avg}} x + \frac{J}{b} x\right] c_{avg} b$$
    $$= \left[3 + \frac{112}{1700}(40) + \frac{0.5}{4}(40)\right](1700)(4)$$
    $$= 72{,}321 \text{ lb/ft}$$
    $$= 6{,}027 \text{ lb/in.}$$

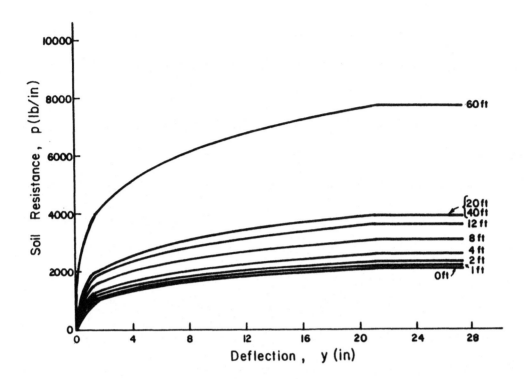

Fig. 31. Example p-y curves for stiff clay above water table, Welch criteria, cyclic loading

From Eq. 2, p_u = 9cb
= (9)(13)(4) = 46,800 lb/ft
= 3,900 lb/in.

Select p_u = 3,900 lb/in.

3. Compute y_{50}:
From Eq. 3, $y_{50} = 2.5\varepsilon_{50}b = (2.5)(0.005)(48)$
= 0.6 in.

4. Compute the points describing the static p-y curve:

From Eq. 20, $\dfrac{p}{p_u} = 0.5\left(\dfrac{y_s}{y_{50}}\right)^{0.25}$ $0 < y_s < 16y_{50}$

$p = p_u$ $y_s \geq 16y_{50}$

Now, $p = 0.5p_u\left(\dfrac{y_s}{y_{50}}\right)^{0.25}$

$= (0.5)(3900)\left(\dfrac{y_s}{y_{50}}\right)^{0.25}$

$= (2215.6)y_s^{0.25}$

5. Compute the points describing the cyclic p-y curve:

From Eq. 21, $\quad C = 9.6\left(\dfrac{p}{p_u}\right)^4$

For p = 1000 $\quad C = 9.6\left(\dfrac{1000}{3900}\right)^4$
$\quad\quad\quad\quad\quad\quad = 0.0415$

Rearranging Eq. 20, $\quad y_s = 16 y_{50}\left(\dfrac{p}{p_u}\right)^4$

$\quad\quad\quad\quad\quad\quad y_s = 16(0.6)\left(\dfrac{1000}{3900}\right)^4$
$\quad\quad\quad\quad\quad\quad y_s = 0.0415 \text{ in}$

From Eq. 22, $\quad y_c = y_s + (y_{50})(C)(\log N)$
$\quad\quad\quad\quad\quad\quad y_c = 0.0415 + (0.6)(0.0415)(\log 100)$
$\quad\quad\quad\quad\quad\quad y_c = 0.0913$

p (lb)	C	y_s (in)	$y_c = y_s + y_{50} \cdot C \cdot \log N$ (in)
1000	0.0415	0.04	0.0913
2000	0.6639	0.664	1.461
3000	3.361	3.361	7.394
3900	9.60	9.60	21.12

An example of the computation of p-y curves for stiff clay above the water table under static loading is shown in Appendix 7.

2.7 p-y CURVES FOR CLAY BELOW WATER TABLE BY UNIFIED METHOD

Introduction

As was noted in the previous sections, no recommendations were made for ascertaining for what range of undrained shear strength one should employ the criteria for soft clay and for what range one should employ the criteria for stiff clay. Sullivan (1977) examined the original experiments and developed a set of recommendations that yield computed behaviors in reasonably good agreement with the experimental results from the Sabine tests reported by Matlock (1970) and with those from the Manor tests reported by Reese, Cox and Koop (1975). However, as will be seen from the following presentation, there is a need for the engineer to employ some judgement in selecting appropriate parameters for use in the prediction equations.

Recommendations for Computing p-y Curves

The following procedure is for <u>short-term static loading</u> and is illustrated in Fig. 32.

1. Obtain values for the undrained shear strength c, the submerged unit weight γ', and the pile diameter b. Also obtain values of ε_{50} from stress-strain curves. If no stress-strain curves are available, the values in Table 4 are provided as guidelines for selection of ε_{50}.

2. Compute c_a and $\bar{\sigma}_v$, for x < 12b, where
    c_a = average undrained shear strength,
    $\bar{\sigma}_v$ = average effective stress, and
    x = depth.

3. Compute the variation of p_u with depth using the equations below.

    a. For x < 12b, p_u is the smaller of the values computed from the two equations below.

    $$p_u = \left(2 + \frac{\bar{\sigma}_v}{c_a} + 0.833 \frac{x}{b}\right) c_a b \qquad (23)$$

    $$p_u = \left(3 + 0.5 \frac{x}{b}\right) c_b \qquad (24)$$

    b. For x > 12b,
    $$p_u = 9cb. \qquad (25)$$

The steps below are for a particular depth, x.

4. Select the coefficients, A and F, as indicated below. The coefficients A and F, determined empirically for the load tests at Sabine and Manor, are given in Table 5. The terms used in Table 5, not defined previously, are defined below.

    W_L = liquid limit,
    PI = plasticity index,
    LI = liquidity index,
    O_R = overconsolidation ratio, and
    S_t = sensitivity.

The recommended procedure for estimating A and F for other clays is given below.

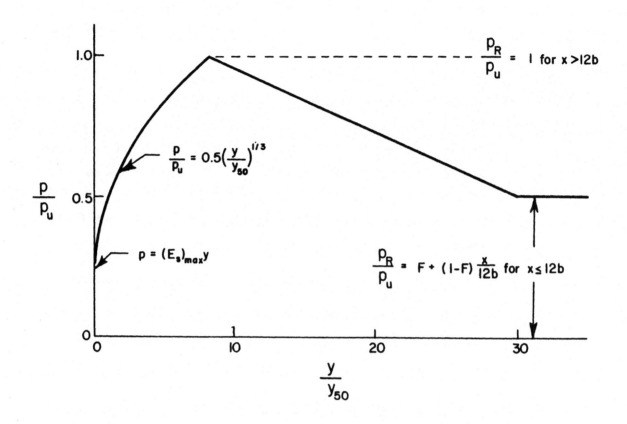

Fig. 32. Characteristic shape of p-y curve for unified clay criteria - static loading

TABLE 4. REPRESENTATIVE VALUES FOR ε_{50}

$\dfrac{c}{(lb/sq\ ft)}$	$\dfrac{\varepsilon_{50}}{\%}$
250 - 500	2
500 - 1000	1
1000 - 2000	0.7
2000 - 4000	0.5
4000 - 8000	0.4

(Also see Tables 3 and 5)

TABLE 5. CURVE PARAMETERS FOR THE UNIFIED CRITERIA

Clay Description	A	F
Sabine River	2.5	1.0
Inorganic, Intact		
$c = 300$ lb/sq ft		
$\varepsilon_{50} = 0.7\%$		
$O_R \approx 1$		
$S_t \approx 2$		
$W_L = 92$		
$PI = 68$		
$LI = 1$		
Manor	0.35	0.5
Inorganic, Very fissured		
$c \approx 2400$ lb/sq ft		
$\varepsilon_{50} = 0.5\%$		
$O_R > 10$		
$S_t \approx 1$		
$W_L = 77$		
$PI = 60$		
$LI = 0.2$		

a. Determine as many of the following properties of the the clay as possible, c, ε_{50}, O_R, S_t, degree of fissuring, ratio of residual to peak undrained shear strength, W_L, PI, and LI.

b. Compare the properties of the soil in question to the properties of the Sabine and Manor clays listed in Table 5.

c. If the properties are similar to either the Sabine or Manor clay properties, use A and F for the similar clay.

d. If the properties are not similar to either, the engineer should estimate A and F using his judgement and Table 5 as guides.

5. Compute:

$$y_{50} = A\varepsilon_{50}b. \quad (26)$$

6. Obtain $(E_s)_{max}$. When no other method is available, Equation 27 and Table 6 may be used as guidelines.

$$(E_s)_{max} = kx \quad (27)$$

TABLE 6. REPRESENTATIVE VALUES FOR k

c	k
(lb/sq ft)	(lb/cu in)
250 - 500	30
500 - 1000	100
1000 - 2000	300
2000 - 4000	1000
4000 - 8000	3000

(Also see Table 2)

7. Compute the deflection at the intersection between the initial linear portion and curved portion, from the equation below.

$$y_g = \left| \frac{0.5 \, p_u}{(E_s)_{max}} \right|^{1.5} (y_{50})^{-0.5} \qquad (28)$$

(y_g can be no larger than $8y_{50}$)

8a. For $0 < y < y_g$,
$$p = (E_s)_{max} y. \qquad (29)$$

8b. For $y_g < y < 8y_{50}$,
$$p = 0.5 \, p_u \left(\frac{y}{y_{50}} \right)^{\frac{1}{3}} \qquad (30)$$

8c. For $8y_{50} < y < 30y_{50}$,
$$p = p_u + \frac{p_R - p_u}{22 \, y_{50}} (y - 8y_{50}) \qquad (31)$$

where
$$p_R = p_u \left(F + (1 - F) \frac{x}{12 \, b} \right). \qquad (32)$$

(p_R will be equal to or less than p_u)

8d. For $y > 30y_{50}$,
$$p = p_R. \qquad (33)$$

The following procedure is for <u>cyclic loading</u> and is illustrated in Fig. 33.

1. Repeat steps 1 through 8a for static loading.
2. Compute

$$p_{CR} = 0.5 \, p_u \frac{x}{12 \, b} \leq 0.5 \, p_u. \qquad (34)$$

3a. For $y_g < y < y_{50}$,
$$p = 0.5 \, p_u \left(\frac{y}{y_{50}} \right)^{\frac{1}{3}} \qquad (35)$$

3b. For $y_{50} < y < 20y_{50}$,
$$p = 0.5 \, p_u + \frac{p_{CR} - 0.5 \, p_u}{19 \, y_{50}} (y - y_{50}). \qquad (36)$$

3c. For $y > 20y_{50}$,
$$p = p_{CR}. \qquad (37)$$

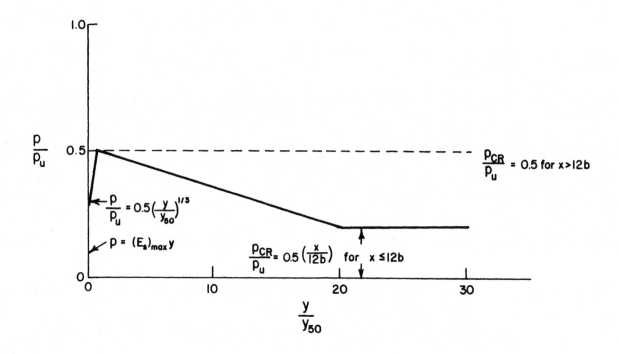

Fig. 33. Characteristic shape of p-y curve for unified clay criteria - cyclic loading

The procedure outlined above for both static and cyclic loading assumes an intersection of the curves defined by Eqs. 29 and 30. If that intersection does not occur, the p-y curve is defined by Eq. 29 until it intersects a portion of the curve defined by Eqs. 31 or 33 for static loading, and Eqs. 35 or 36 for cyclic loading.

Example Curves

Two example sets of p-y curves were computed using the unified criteria; each of the sets is for a pile of 48 inches in diameter and for cyclic loading.

Figure 34 shows the set of p-y curves for soft clay; the soil profile used is shown in Fig. 23. The value of ε_{50} was assumed to be 0.02 at the mudline and 0.01 at a depth of 80 ft. The unit weight was assumed to be 20 lb/cu ft at the groundline and 40 lb/cu ft at a depth of 80 ft. The value of A was assumed to be 2.5 and the value of F was assumed to be 1.0. The value of k for computing the maximum value of the soil modulus was assumed to be 231 lb/cu in. The p-y curves were computed for the following depths: 0, 1, 2, 4, 8, 12, 20, and 40 ft. (Curves for 0 and 1 ft too close to axis to be shown.)

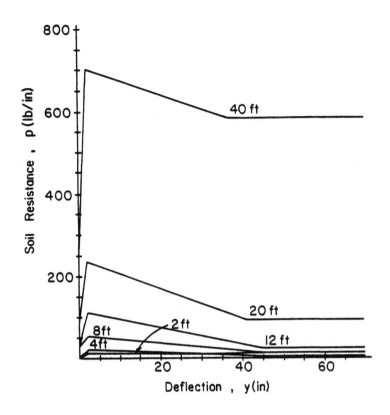

Fig. 34. Example p-y curves for soft clay below water table, unified criteria, cyclic loading

Figure 35 shows the set of p-y curves for stiff clay; the soil profile used is shown in Fig. 28. The value of ε_{50} was assumed to be 0.006 and the unit weight of the soil was assumed to be 50 lb/cu ft. The value of A was assumed to be 0.35 and the value of F was assumed to be 0.5. The value of k for computing the maximum value of the soil modulus was assumed to be 463 lb/cu in. The p-y curves were computed for the following depths: 0, 1, 2, 4, 8, 12, 20, and 40 ft.

2.8 p-y CURVES FOR SAND

As shown below, a major experimental program was conducted on the behavior of laterally loaded piles in sand below the water table. The results can be extended to sand above the water table by making appropriate adjustments in the values of the unit weight, depending on the position of the water table.

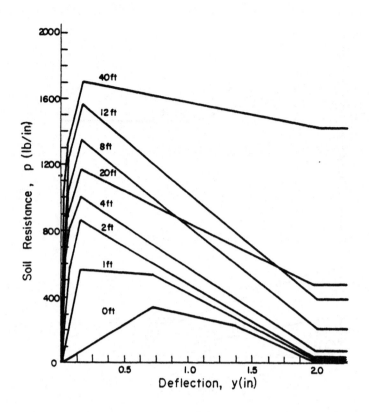

Fig. 35. Example p-y curves for stiff clay below water table, unified criteria, cyclic loading

Field Experiments

An extensive series of tests were performed at a site on Mustang Island, near Corpus Christi (Cox, Reese, and Grubbs, 1974). Two steel-pipe piles, 24 inches in diameter, were driven into sand in a manner to simulate the driving of an open-ended pipe, and were subjected to lateral loading. The embedded length of the piles was 69 ft. One of the piles was subjected to short-term loading and the other to repeated loading.

The soil at the site was a uniformly graded, fine sand with an angle of internal friction of 39 degrees. The submerged unit weight was 66 lb/cu ft. The water surface was maintained a few inches above the mudline throughout the test program.

The results from the field experiments led to the recommendations shown in the following paragraphs.

Recommendations for Computing p-y Curves

The following procedure is for <u>short-term static loading</u> and for <u>cyclic loading</u> and is illustrated in Fig. 36 (Reese, Cox, and Koop, 1974).

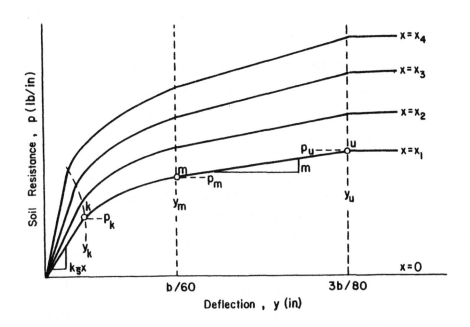

Fig. 36. Characteristic shape of a family of p-y curves for static and cyclic loading in sand

1. Obtain values for the angle of internal friction ϕ, the soil unit weight γ, and pile diameter b.

2. Make the following preliminary computations.

$$\alpha = \frac{\phi}{2}; \quad \beta = 45 + \frac{\phi}{2}; \quad K_o = 0.4; \quad \text{and} \quad K_a = \tan^2(45 - \frac{\phi}{2}) \quad (38)$$

$$K_p = \tan^2(45 + \frac{\phi}{2})$$

3. Compute the ultimate soil resistance per unit length of pile using the smaller of the values given by the equations below, where x is equal to the depth below the ground surface (Fenske, 1981).

$$p_{st} = \gamma b^2 [S_1(x/b) + S_2(x/b)^2] \quad (39)$$

$$p_{sd} = \gamma b^2 [S_3(x/b)] \quad (40)$$

where

$$S_1 = (K_p - K_a) \quad (41)$$

$$S_2 = (\tan \beta)(K_p \tan \alpha + K_o[\tan \phi \sin \beta(\sec \alpha + 1) - \tan \alpha]) \tag{42}$$

$$S_3 = K_p^2(K_p + K_o \tan \phi) - K_a. \tag{43}$$

4. The depth of transition x_t can be found by equating the expressions in Eqs. 39 and 40, as follows:

$$x_t/b = (S_3 - S_1)/S_2. \tag{44}$$

The appropriate γ for the position of the water table should be employed. Use Eq. 39 above x_t and Eq. 40 below. It can be seen that S_1, S_2, S_3, and x_t/b are functions only of ϕ; therefore, the values shown in Table 7 can be computed.

TABLE 7. NONDIMENSIONAL COEFFICIENTS FOR p-y CURVES FOR SAND (after Fenske)

ϕ, deg.	S_1	S_2	S_3	x_t/b
25.0	2.05805	1.21808	15.68459	11.18690
26.0	2.17061	1.33495	17.68745	11.62351
27.0	2.28742	1.46177	19.95332	12.08526
28.0	2.40879	1.59947	22.52060	12.57407
29.0	2.53509	1.74906	25.43390	13.09204
30.0	2.66667	1.91170	28.74513	13.64147
31.0	2.80394	2.08866	32.51489	14.22489
32.0	2.94733	2.28134	36.81400	14.84507
33.0	3.09732	2.49133	41.72552	15.50508
34.0	3.25442	2.72037	47.34702	16.20830
35.0	3.41918	2.97045	53.79347	16.95848
36.0	3.59222	3.24376	61.20067	17.75976
37.0	3.77421	3.54280	69.72952	18.61673
38.0	3.96586	3.87034	79.57113	19.53452
39.0	4.16799	4.22954	90.95327	20.51883
40.0	4.38147	4.62396	104.14818	21.56704

5. Select a depth at which a p-y curve is desired.

6. Establish y_u as $3b/80$. Compute p_u by the following equation:
$$p_u = \bar{A}_s p_s \quad \text{or} \quad p_u = \bar{A}_c p_s. \tag{45}$$
Use the appropriate value of $\bar{A}_s$ or $\bar{A}_c$ from Fig. 37 for the particular nondimensional depth, and for either the static or cyclic case. Use the appropriate equation for p_s, Eq. 39 or Eq. 40 by referring to the computation in step 4.

7. Establish y_m as $b/60$. Compute p_m by the following equation:
$$p_m = B_s p_s \quad \text{or} \quad p_m = B_c p_s. \tag{46}$$
Use the appropriate value of B_s or B_c from Fig. 38 for the particular nondimensional depth, and for either the static or cyclic case. Use the appropriate equation for p_s. The two straight-line portions of the p-y curve, beyond the point where y is equal to $b/60$, can now be established.

8. Establish the initial straight-line portion of the p-y curve,
$$p = (kx)y. \tag{47}$$
Use Tables 8 and 9 to select an appropriate value of k.

9. Establish the parabolic section of the p-y curve,
$$p = \bar{C} y^{1/n}. \tag{48}$$
Fit the parabola between points k and m as follows:

a. Get the slope of line between points m and u by,
$$m = \frac{p_u - p_m}{y_u - y_m}. \tag{49}$$

b. Obtain the power of the parabolic section by,
$$n = \frac{p_m}{m y_m}. \tag{50}$$

c. Obtain the coefficient $\bar{C}$ as follows:
$$\bar{C} = \frac{p_m}{y_m^{1/n}}. \tag{51}$$

d. Determine point k as,
$$y_k = \left(\frac{\bar{C}}{kx}\right)^{n/n-1}. \tag{52}$$

e. Compute appropriate number of points on the parabola by using Eq. 48.

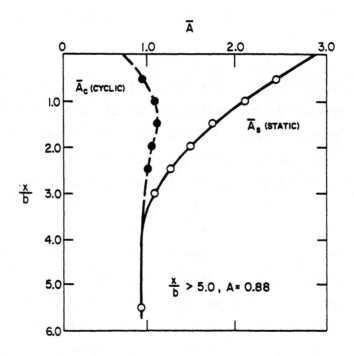

Fig. 37. Values of coefficients $\bar{A}_c$ and $\bar{A}_s$

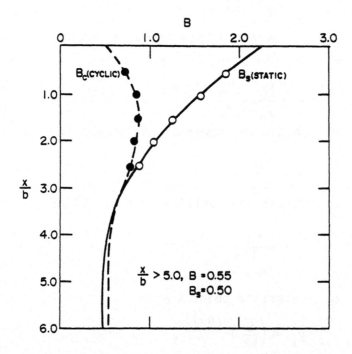

Fig. 38. Nondimensional coefficient B for soil resistance versus depth

TABLE 8. RELATIONSHIP BETWEEN N AND ϕ AND D_r
(after Gibbs and Holtz, 1957)

N(SPT) blows/ft	Overburden Stress, lb/sq in					
	0		20		40	
	ϕ degrees	D_r percent	ϕ degrees	D_r percent	ϕ degrees	D_r percent
0	0	0	0	0	0	0
2	32	45				
4	34	55				
6	36	65	30	37		
10	38	75	32	46	31	40
15	42	90	34	57	32	48
20	45	100	36	65	34	55
25			37	72	35	60
30			39	77	36	65
35			40	82	36	67
40			41	86	37	72
45			42	90	38	75
50			44	95	39	77
55			45	100	39	80
60					40	83
65					41	86
70					42	90
75					42	92
80					43	95
85					44	97
90					44	99

D_r = relative density; ϕ = angle of internal friction

N = blow count from standard penetration test

Note: The step-by-step procedure is outlined, and Fig. 36 is drawn, as if there is an intersection between the initial straight-line portion of the p-y curve and the parabolic portion of the curve at point k. However, in some instances there may be no intersection with the parabola. Equation 47 defines the p-y curve until there is an intersection with another branch of the p-y curve or if no intersection occurs, Eq. 47 defines the complete p-y curve. The soil-response curves for other depths can be found repeating the above steps for each desired depth.

TABLE 9. REPRESENTATIVE VALUES OF k (lb/cu in) FOR SAND

Relative Density	below 35%	35% to 65%	above 65%
Recommended k for sand below water table	20	60	125
Recommended k for sand above water table	25	90	225

Recommended Soil Tests

Triaxial compression tests are recommended for obtaining the angle of internal friction of the sand. Confining pressures should be used which are close or equal to those at the depths being considered in the analysis. Tests must be performed to determine the unit weight of the sand. In many instances, however, undisturbed samples of sand cannot be obtained and the value of ϕ must be obtained from correlations with static cone penetration tests or from dynamic penetration tests (see Table 8).

Example Curves

An example set of p-y curves was computed for sand below the water table for a pile with a diameter of 48 in. The sand is assumed to have an angle of internal friction of 34° and a submerged unit weight of 62.4 lb/cu ft. The loading was assumed to be cyclic.

The p-y curves were computed for the following depths below the mudline: 0, 1, 2, 4, 8, 12, and 20 ft. The plotted curves are shown in Fig. 39.

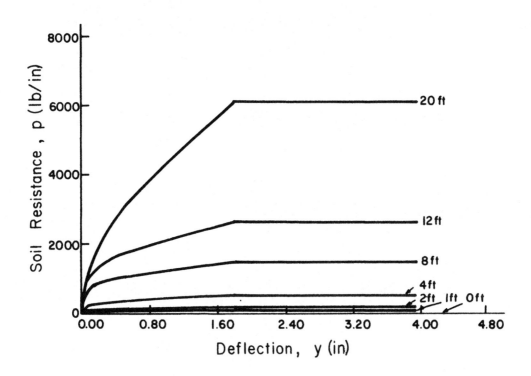

Fig. 39. Example p-y curves for sand below water table, Reese criteria, cyclic loading

The step-by-step solutions for computing the cyclic p-y curve at a depth of 8 ft are shown as follows:

1. Obtain the best possible estimate of soil parameters:

    at depth 8 ft: $\phi = 34°$

    $\gamma = 62.4$ lb/cu ft

    $k = 60$ lb/cu in (from Table 9)

    $b = 48$ in.

2. Make the following preliminary computations:

    $\alpha = \phi/2 = 17°$, $\beta = 45 + \phi/2 = 62°$, $K_o = 0.4$

    $K_a = \tan^2(45° - \phi/2) = 0.283$ (Eq. 38)

    $K_p = \tan^2(45° + \phi/2) = 3.537$ (Eq. 38)

3. Compute the ultimate soil resistance per unit length of pile:

From Eq. 39, $p_{st} = \gamma b^2[S_1(x/b) + S_2(x/b)^2]$

From Eq. 40, $p_{sd} = \gamma b^2[S_3(x/b)]$

From Table 7, $S_1 = 3.25442$, $S_2 = 2.72037$, $S_3 = 47.34702$ were selected to compute p_{st} and p_{sd}.

$p_{st} = (62.4)(4^2)[3.25442(8/4) + 2.72037(8/4)^2]$
$= 17362 \text{ lb/ft} = 1447 \text{ lb/in}$

$p_{sd} = (62.4)(4^2)(47.34702)(8/4) = 94543 \text{ lb/ft}$
$= 7879 \text{ lb/in}.$

Select $p_s = 1447$ lb/in.

4. Compute transition depth x_t:

From Table 7, $x_t/b = 16.20830$ was selected.

5. (The depth of 8 ft was previously selected.)

6. Compute p_u at $y_u = 3b/80$:

$y_u = 3b/80 = \dfrac{(3)(48)}{80} = 1.8$ in.

From Eq. 45, $p_u = A_c p_s$

From Fig. 37, $A_c = 1.03$

So, $p_u = \bar{A}_c p_s$
$= (1.03)(1447) = 1490$ lb/in.

7. Compute p_m at $y_m = b/60$:

$y_m = b/60 = 48/60 = 0.8$ in.

From Eq. 46, $p_m = B_c p_s$

From Fig. 38, $B_c = 0.82$

So, $p_m = B_c p_s = (0.82)(1447) = 1186$ lb/in

8. Establish the initial straight-line portion of the p-y curve:

From Eq. 47, $p = (kx)y$
$= (60)(96)y$
$= 5760y$

9. Establish the parabolic section of the p-y curve:

From Eq. 48, $p = \bar{C} y^{1/n}$

a. From Eq. 49, $m = \dfrac{p_u - p_m}{y_u - y_m} = \dfrac{1490 - 1186}{1.8 - 0.8} = 304$

b. From Eq. 50, $n = \dfrac{p_m}{m y_m} = \dfrac{1186}{(304)(0.8)} = 4.877$

c. From Eq. 51, $\bar{C} = \dfrac{p_m}{y_m^{1/n}} = \dfrac{1186}{0.8^{1/4.877}} = 1242$

then $p = \bar{C} y^{1/n}$

$= 1242 y^{1/4.877} = 1242 y^{0.205}$

d. From Eq. 52, $y_k = \left(\dfrac{\bar{C}}{kx}\right)^{n/n-1} = \left(\dfrac{1242}{(60)(96)}\right)^{\frac{4.877}{3.877}}$

$= 0.1452$ in

$p_k = (kx)y_k = (60)(96)(0.1452)$
$= 836$ lb/in.

e. Compute appropriate number of points on the parabola by using Eq. (48):

$p = \bar{C}^{1/n}$
$= 1242 y^{0.205}$

y (in)	p (lb/in)
0.145	836
0.200	894
0.400	1031
0.600	1120
0.700	1156

2.9 p-y CURVES FOR ROCK

It is hardly surprising that not much information is available on the behavior of piles that have been installed in rock. Some other type of foundation would normally be used. However, a study was made of the behavior of an instrumented drilled shaft that was installed in vuggy limestone in the Florida Keys (Reese and Nyman, 1978). The test was performed for the purpose of gaining information for the design of foundations for highway bridges in the Florida Keys.

Difficulty was encountered in obtaining properties of the intact rock. Cores broke during excavation and penetrometer tests were misleading (because of the vugs) or could not be run. It was possible to test two cores from the site. The small discontinuities on the outside surface of the specimens were covered with a thin layer of gypsum cement in an effort to minimize stress concentrations. The ends of the specimens were cut with a rock saw and lapped flat and parallel. The specimens were 5.88 inches in diameter and with heights of 11.88 in for Specimen 1 and 10.44 in for Specimen 2. The undrained shear strength of the specimens were taken as one-half the unconfined compressive strength and were 17.4 and 13.6 T/sq ft for Specimens 1 and 2, respectively.

The rock at the site was also investigated by in-situ-grout-plug tests under the direction of Dr. John Schmertmann (1977). A 5.5-in diameter hole was drilled into the limestone, a high strength steel bar was placed to the bottom of the hole, and a grout plug was cast over the lower end of the bar. The bar was pulled until failure occurred and the grout was examined to see that failure occurred at the interface of the grout and limestone. Tests were performed at three borings and the following results were obtained, in T/sq ft: depth into limestone from 2.5 to 5 ft, 23.8, 13.7, and 12.0; depth into limestone from 8 to 10 ft, 18.2, 21.7, and 26.5; depth into limestone from 18 to 20 ft, 13.7 and 10.7. The average of the eight tests was 16.3 T/sq ft. However, the rock was stronger in the zone where the deflections of the drilled shaft were most significant and a shear strength of 18 T/sq ft was selected for correlation.

The drilled shaft was 48 inches in diameter and penetrated 43.7 ft into the limestone. The overburden of fill was 14 ft thick and was cased. The load was applied about 11.5 ft above the limestone. A maximum load of 75 tons was applied to the drilled shaft. The maximum deflection at the point of load application was 0.71 inch and at the top of the rock (bottom of casing) it was 0.0213 in. While the curve of load versus deflection was nonlinear, there was no indication of failure of the rock.

A single p-y curve, shown in Fig. 40, was proposed for the design of piles under lateral loading in the Florida Keys. Data are insufficient to indicate a family of curves to reflect any increased resistance with depth. Cyclic loading caused no measurable decrease in resistance by the rock.

As shown in the figure, load tests are recommended if deflections of the rock (and pile) are greater than 0.004b and brittle fracture is assumed if the lateral stress (force per unit length) against the rock becomes greater than the diameter times the shear strength s_u of the rock.

The p-y curve shown in Fig. 40 should be employed with considerable caution because of the limited amount of experimental data and because of the great variability in rock. The behavior of rock at a site could very well be controlled not by the strength of intact specimens but by joints, cracks, and secondary structure of the rock.

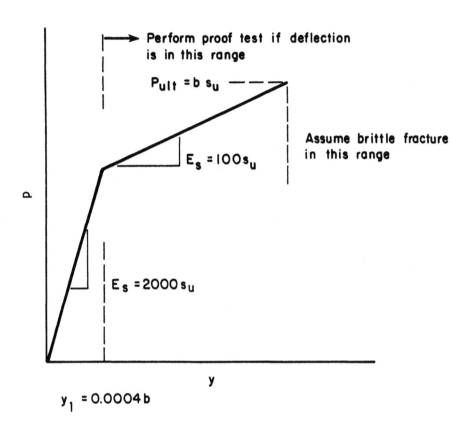

Fig. 40. Recommended p-y curve for design of drilled shafts in vuggy limestone

CHAPTER 3. COMPUTER METHOD OF ANALYSIS

3.1 ANALYTICAL METHOD

The solution of the problem of the pile under lateral load must satisfy two general conditions. The equations of equilibrium must be solved and deflections and deformations must be consistent and compatible. These two requirements are fulfilled by finding a solution to the following differential equation (Hetenyi, 1946).

$$EI \frac{d^4y}{dx^4} + P_x \frac{d^2y}{dx^2} - p - W = 0 \tag{53}$$

where

P_x = axial load on the pile,
y = lateral deflection of the pile at a point x along the length of the pile,
p = soil reaction per unit length,
EI = flexural rigidity, and
W = distributed load along the length of the pile.

Other beam formulae which are useful in the analysis are:

$$EI \frac{d^3y}{dx^3} = V, \tag{54}$$

$$EI \frac{d^2y}{dx^2} = M, \tag{55}$$

and,

$$\frac{dy}{dx} = S, \tag{56}$$

where

V = shear,
M = bending moment of the pile, and
S = slope of the elastic curve.

The derivation of the above equations is discussed in Chapter 2 (Reese, 1983).

Solutions of the above equations can be made by use of the computer program described in this chapter. Nondimensional methods, described later, can frequently be used to obtain acceptable solutions but those methods are much less versatile than the computer method.

An acceptable technique for getting solutions to the equations governing the behavior of a laterally loaded pile is to formulate the differential equation in difference terms. The pile is divided into n increments of constant length h. Equation 53 can be represented at point m along the pile as follows:

$$y_{m-2}R_{m-1} + y_{m-1}(-2R_{m-1} - 2R_m + P_x h^2)$$
$$+ y_m(R_{m-1} + 4R_m + R_{m+1} - 2P_x h^2 + k_m h^4)$$
$$+ y_{m+1}(-2R_m - 2R_{m+1} + P_x h^2) + y_{m+2}R_{m+1} - W_m = 0 \quad (57)$$

where

y_m = deflection at point m,
$R_m = E_m I_m$ = flexural rigidity at point m,
P_x = axial load (causes no moment at x = 0),
$k_m = p_m/y_m$ = soil modulus at point m, and
W_m = distributed load at point m.

Because the pile was divided into n increments, there are n+1 points on the pile and n+1 of the above equations can be written. It can be seen, however, that the differential equation in difference form uses deflections at two points above and at two points below the point being considered. Therefore, four imaginary deflections are introduced, two at the top of the pile and two at the bottom. The introduction of four boundary conditions, two at the bottom of the pile and two at the top, yields n+5 simultaneous equations of a sort to be easily and quickly solved by the digital computer.

After deflections are found by solving the simultaneous equations, shear, moment, and slope can be found at all points along the pile by solving Eqs. 54, 55, and 56. The soil resistance p can be found by the product $k_m y_m$. It is obvious that an iterative solution must be made with the computer because the values of the soil moduli k_m are not known at the outset. The computer is programmed to begin with a set of deflections as a trial and then values of soil modulus and deflections are successively computed until convergence is achieved. Convergence to the correct solution is judged to have been achieved when the differences between the final two sets of computed deflections are less than the value of the tolerance selected by the engineer. Suggestions on selecting values of tolerance are given later.

3.2 BOUNDARY CONDITIONS

At the bottom of the pile the two boundary conditions employed are the shear and the moment and both are equal to zero. Thus, a solution can be obtained for a short pile such that there is a significant amount of deflection and slope at the bottom of the pile. Sometimes the question arises about the possibility of forces at the base of the pile due to development of shearing stresses from the soil when the bottom of the pile is deflected. That possibility can readily be accomodated by placing a p-y curve with appropriate numerical values at the bottom increment of the pile.

Strictly speaking, there are three boundary conditions to be selected at the top of the pile but one of those, the axial load, does not provide any specific information on pile-head deflection. Thus, two other boundary conditions must be selected. The computer is programmed to accept one of the following three sets. (The axial load is assumed to be used with each of these sets).

1. The lateral load (P_t) and the moment (M_t) at the top of the pile are known.
2. The lateral load (P_t) and the slope of the elastic curve (S_t) at the top of the pile are known.
3. The lateral load (P_t) and the rotational-restraint constant (M_t/S_t) at the top of the pile are known.

The first set of boundary conditions applies to a case such as a highway sign where wind pressure applies a force some distance above the groundline. The axial load will usually be small and a free body of the pile can be taken at the groundline where the shear and the moment will be known. The second set of boundary conditions can be employed if a pile supports a retaining wall or bridge abutment and where the top of the pile penetrates some distance into a reinforced concrete mat. The shear will be known and the pile head rotation in most cases can be assumed to be zero. The third set of boundary conditions is encountered when a pile frames into a super-structure that is flexible. In some bridge structures the piles could continue and form the lower portion of a column. A free body of the pile can be taken at a convenient point and the rotational restraint (M_t/S_t) of the portion of the structure above the pile head can be estimated. The magnitude of the shear will be known. Iteration between pile and superstructure will lead to improved values of rotational restraint and convergence to an appropriate solution can be achieved.

3.3 COMPUTER PROGRAM COM624

The method for solving the governing equations and the recommendations for p-y curves have been incorporated into a computer program that is available from the Geotechnical Engineering Center of The University of Texas at Austin (See Appendix 5.) A nominal charge is made for the program and the associated documentation and a users' group is established. The program and documentation are copyrighted in order to ensure that the location of each program is known and so that any new information on the best use of the program can be immediately sent to all the users. The technology is so straightforward, however, that some engineering offices may prefer to write their own program.

The following presentation in this section is based on the assumption that COM624 is available. All of the information given here is presented in the documentation of COM624, along with additional material, but the partial repetition here allows for completeness and clarity.

3.4 STEP-BY-STEP PROCEDURE OF ANALYSIS

The detailed procedure for the analysis of a single pile under lateral load by computer can be illustrated by examining the guide for the computer input. Accordingly, Appendix 1 has been provided. Sections A1.1 through A1.5 give exact information on the use of the computer program. The material in Appendix 1 is self-explanatory; however, a few comments are made below in further explanation. The step-by-step procedure is illustrated in the solution of example problems shown in the next section.

The selection of units in making computations is at the the option of the user. However, the input and the output are in consistent units. For example, if English units are selected, the input and output are in pounds and inches. The slight inconvenience in not using kips and feet is more than offset by the elimination of the potential for errors. Of course, kips and feet can be used if the engineer decides to convert all input into those terms. The pile stiffness (EI) would then be in unfamiliar units.

Two factors influence the accuracy of the computer solution: the word length employed by the computer, and the increment length h.

Equation 57 can be solved with appropriate accuracy, assuming the value of h has been selected properly, if the computer has a single-precision word length of about 60 bits. Double precision is necessary if the word length is in the order of 30 to 35 bits; a double-precision version of COM624 is available.

The selection of the increment length must be made in consideration of the stiffness of the pile. An increment length equal to about one-half the pile diameter has been found frequently to be acceptable. The example problem illustrates the effect of changing the length of the increment.

A discussion was presented earlier of the importance of the length of the pile. If a pile sustains only lateral load, a length should be selected so that a soil failure is avoided. However, most of the piles encountered in practice are subjected both to axial and to lateral loading. The axial loading frequently controls the length of the pile. If such is the case, the computer output for lateral loading will show several points of zero deflection. After a length exhibiting about two points of zero deflection the pile behaves like a pile of infinite length; therefore, in the computer analysis the length can be arbitrarily reduced in order to reduce running time and to make the output more readable.

3.5 EXAMPLE PROBLEM, STEEL PILE SUPPORTING BRIDGE ABUTMENT

The example that follows is intended to demonstrate the capability of the computer program and is not intended to be a complete presentation of the design process. There are, of course, a number of factors beyond those given here that must be considered in design; however, the example should serve to give a helpful step-by-step approach in the use of COM624.

Pile

14HP89, depth = 13.83 in, width = 14.70 in, I_x = 904 in^4, t_w = 0.615 in, Z_x (plastic modulus) = 146 cu in, F_y = 40 kips/sq in, length = 600 in (50 ft).

Soil

Stiff clay above the water table, γ = 0.069 lb/cu in (119 lb/cu ft), c = 14.0 lb/sq in (2016 lb/sq ft), ε_{50} = 0.007.

Application and Loading of Foundation for Bridge Abutment or for a Retaining Wall

It is assumed that the design axial load is 50 kips (the service load times the factor of safety) and it is desired to find the lateral load that will cause a soil failure (excessive deflection) or a pile failure (development of plastic hinge). The pile head is assumed to be fixed against rotation. It is further assumed that the sustained load from earth pressures will not cause consolidation or creep of the overconsolidated clay.

Computation of Yield Moment (Horne, 1978)

With no axial load, bending about strong axis

$$M_y = F_y Z_x = (40)(146) = 5840 \text{ in-kips.}$$

Considering effect of axial load

$$a = \frac{P_x}{2t_w F_b} = \frac{50}{(2)(0.615)(40)} = 1.016 \text{ in}$$

$$M_y = M_y - t_w a^2 F_y = 5840 - (0.615)(1.016)^2(40) = 5814 \text{ in-kips.}$$

Design Approach

A number of approaches may be used in the design of a pile under lateral (and axial) loading. The approach that is illustrated here is to find the loading that will produce a plastic hinge (the ultimate bending moment M_p) and to reduce the loading by an appropriate factor of safety. For the example that is presented, an overall factor of safety of 2.5 is employed. The axial load under service conditions was assumed to be 20 kips so a design axial load of 50 kips is employed in the analysis (see computation of yield moment). The procedure is to employ a series of lateral loads, to plot a curve showing bending moment versus lateral load (with an axial load of 50 kips), and to find the lateral load that produces the yield moment in the pile. That lateral load is divided, then, by the factor of safety to find the service load. The service loads, both lateral and axial, are then used to compute curves showing pile deflection and bending moment as a function of depth. These curves are studied as a further evaluation of the design. If the service lateral load so obtained does not satisfy the loading conditions, another pile is selected and the process is repeated.

Input Data

The data for one of the computer runs are shown in Table 10. The input follows the guide in Appendix 1 and should be self-explanatory. As may be seen, English units were selected and the pile was sub-divided into 120 increments, giving an increment length of 5 in. The problem was run on a Control Data Corporation computer that uses a 64-bit word so single-precision arithmetic was satisfactory. Line 8 was left blank so that

TABLE 10. DATA FOR ONE OF THE COMPUTER RUNS

```
                                 COLUMN NUMBER
LINE   ─────────────────────────────────────────────────────
NO.    ....*....1....*....2....*....3....*....4....*....5....*....6

 1     CASE1 STIFF CLAY ABOVE WATER TABLE H-PILE STATIC LOADING
 2     ENGL
 3       120     1      1     0
 4        2      2      0
 5       6.00E2       2.90E7       0.00E0
 6        1      3
 7        2      1      1
 8
 9       0.00E0       1.47E1       9.04E2       2.61E1
10        1      3      0.00E0       7.00E2       4.00E2
11       0.00E0       6.90E-2
12       7.00E2       6.90E-2
13       0.00E0       1.40E1                    7.00E-3
14       7.00E2       1.40E1                    7.00E-3
15        7
16       0.00E0
17       1.00E1
18       2.00E1
19       4.00E1
20       7.00E1
21       1.00E2
22       1.50E2
23        1     3.60E4       0.00E0       5.00E4
24       -1
25     END
       ....*....1....*....2....*....3....*....4....*....5....*....6
```

the maximum number of iterations that were allowed were 100 (26 were actually used), and the tolerance for convergence (difference between successive deflections) was 1×10^{-5} in. There was no distributed load so no input was required for that loading. In line 9, the diameter of the pile was entered as 14.7 in, corresponding to the width. In line 10, the soil profile was shown to a depth of 700 in, a depth sufficient to cover the full length of the pile. The value of k in the expression $E_s = kx$ was selected as 400 lb/cu in. The value of k is relatively unimportant in the present analysis (stiff clay above the water table) because it is not a part of the soil criteria. However, the value is employed in the first iteration. Some typical values are shown in Table 2. Lines 15 through 22 give instructions for generating p-y curves that are to be printed (see Appendix 2). Such curves are useful to the designer to allow checks of the output. Line 23 includes values of loading at pile head and up to 20 different loads can be input without the necessity of altering other data. As may be seen, for this example run only one card was used, showing a lateral load of 36,000 lb and an axial load of 50,000 lb. The pile head was fixed against rotation. The last two lines are for control cards to stop the reading of data on loading and to stop reading all data. Three other examples for coding input forms are shown in Appendix 7.

Output of Computer Program

An example of the computer output is shown in Appendix 2. The results are self-explanatory. It is of interest to note that there are many points of zero deflection along the length of the pile; therefore, the pile length in the computer runs could have been reduced with no loss of accuracy and for economy.

In addition to deflection and bending moment as a function of depth, the output gives computed values of total stress and the soil modulus. As noted earlier, the input was written to require p-y curves to be computed; these are shown. Also, the output includes the "echo print" of significant input values that enable the engineer to be certain that data were input correctly.

Results of Computer Solutions

The maximum bending moment and the pile-head deflection as a function of lateral load were obtained from additional computer output and are plotted in Fig. 41. As may be seen in Fig. 41a, a lateral load of 98 kips produced the yield moment in the pile. Figure 41b shows that no soil failure occurred at the ultimate load of 98 kips.

Behavior of Pile Under Service Loads

The ultimate lateral load of 98 kips and the ultimate axial load of 50 kips were reduced, by employing the factor of safety, to 39 kips and 20 kips, respectively. A final computer run was made and the detailed input and output are shown in Appendix 2. Curves showing bending moment versus depth and deflection versus depth are presented in Figs. 42 and 43, respectively. As may be seen in Fig. 42, the maximum bending moment occurs at the groundline where the pile head is fixed against rotation. The maximum combined stress is 14.1 kips/sq in. Figure 43 shows that the maximum deflection is 0.14 in. If the combined stress and groundline deflection under the service loads are unsatisfactory, a different section can be selected and the computations repeated.

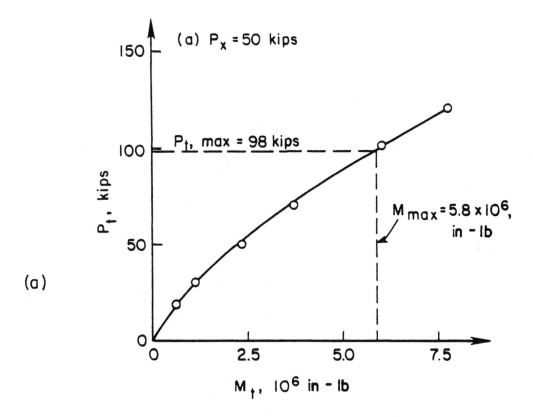

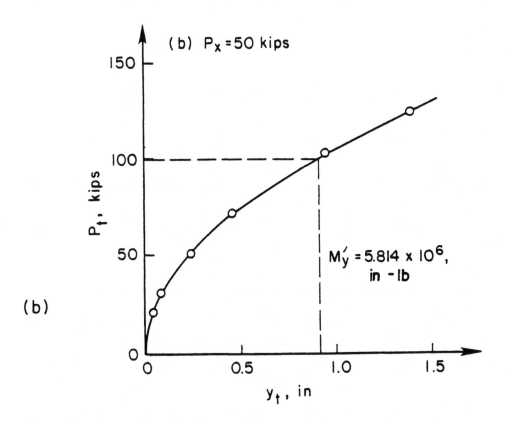

Fig. 41. Results from computer solutions for example problem, steel pile supporting bridge abutment

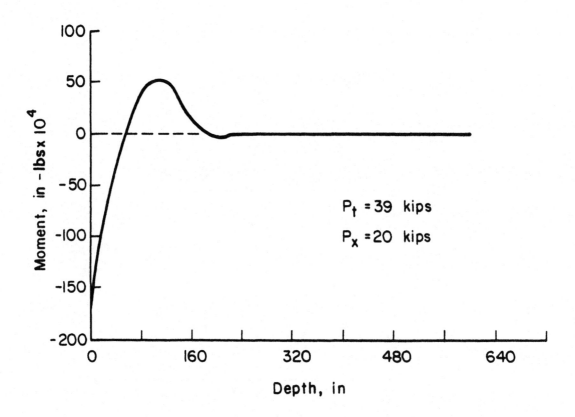

Fig. 42. Bending moment versus depth for example problem, steel pile supporting bridge abutment

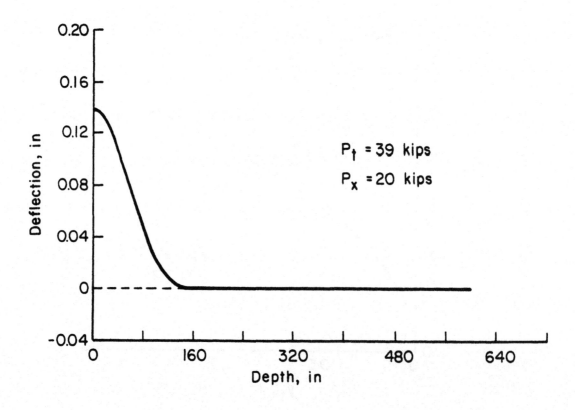

Fig. 43. Deflection versus depth for example problem, steel pile supporting bridge abutment

3.6 EXAMPLE PROBLEM, STEEL PILE SUPPORTING BRIDGE PIER

The example in this section is designed to show the influence of the nature of loading and of the presence of water above the ground surface. The assumption is made that a bridge pier is supported by steel piles like the one in the previous example, that the pier is subjected to cyclic lateral loads from various sources, and that water exists above the ground surface.

The axial load of 50 kips is acting as before and lateral loads are applied to find the maximum bending moment and groundline deflection. The results from the computer runs are shown in Figs. 44a and 44b. When these figures are compared to those in Fig. 41, it can be seen that the load at which the ultimate bending moment is reached is reduced by 49 percent. The deflection at 52 kips was 0.25 in for the abutment problem and was 0.30 in for the pier problem. The detailed input and output for the computer analysis for this example are shown in Appendix 2. The example serves to illustrate the importance of selecting the appropriate loading condition, whether static or cyclic, and of deciding on the probable location of the water table.

3.7 EXAMPLE PROBLEM, DRILLED SHAFT SUPPORTING BRIDGE ABUTMENT

This example is selected to demonstrate the difference in the method of analysis of a drilled shaft as compared to the analysis of a steel section. It would be desirable to select a drilled shaft that is comparable in some way to the steel section; however, that approach is not straightforward. Thus, a drilled shaft is selected that can readily be constructed at the site. All of the other conditions are the same as for the steel section that supports the bridge abutment.

Drilled Shaft

The section is assumed to be 30 in. in outside diameter and to have 12 No. 8 rebars placed on a 24-in diameter circle. The ultimate strength of the concrete is assumed to be 4000 lb/sq in.

Computation of Bending Stiffness and Ultimate Moment

A computer program, PMEIX, has been written to compute EI and M_y and is presented in Appendix 3. (See Appendix 5 for an order blank.) The documentation in the appendix should allow the program to be used without much difficulty. Figure 45 is an interaction diagram obtained by use of

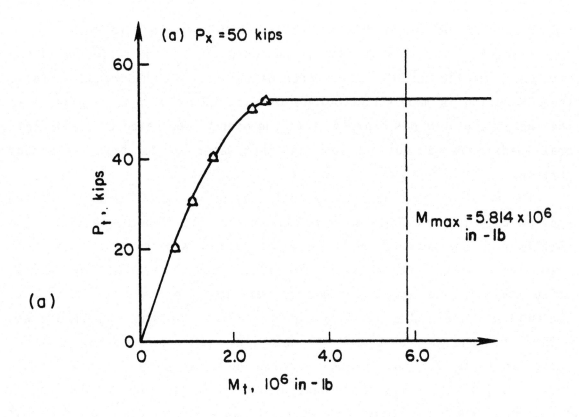

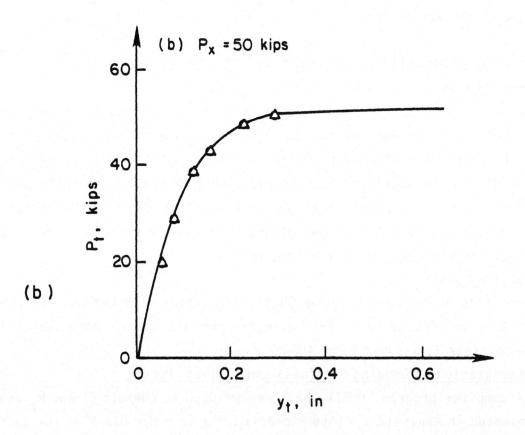

Fig. 44. Results from computer solutions for example problem, steel pile supporting bridge pier

PMEIX. It shows the results of the computations for the ultimate bending moment with the ultimate bending moment being selected at a concrete strain of 0.003. With an axial load of 50 kips, the ultimate bending moment is 6.78 x 10⁶ in-lb. The computations could have been done only with an axial load of 50 kips; however, the interaction diagram is developed in its entirety to show the capability of PMEIX.

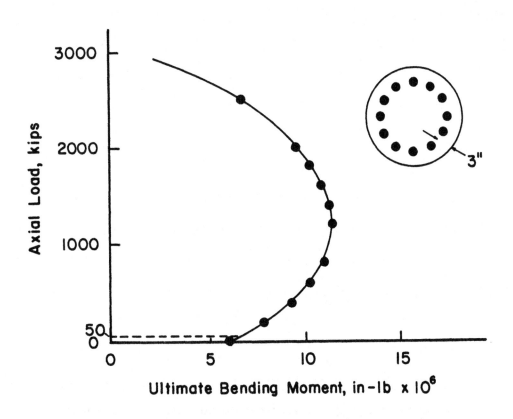

Fig. 45. Interaction diagram for drilled shaft of example problem, from Computer Program PMEIX

With regard to the selection of EI, PMEIX has been used to compute values of EI for an axial load of 50 kips and as a function of applied moment. The results are shown in Table 11. Also shown in Table 11 are the ratios of the applied moment to the ultimate moment. Because only a short portion of the drilled shaft will be subjected to the ultimate moment, a constant EI is selected for further computations. It seems appropriate to select an EI toward the lower range of applied moments. Some studies have indicated that the EI where M_i/M_{ult} is equal to about 13 percent should give results that are approximately correct.

TABLE 11. COMPUTED VALUES OF BENDING STIFFNESS FOR DRILLED SHAFT OF EXAMPLE PROBLEM

EI NO.	$\dfrac{M_i*}{M_{ult}}$	EI at M_i lb-sq in
1	3%	1.89×10^{11}
2	5%	1.43×10^{11}
3	13%	6.96×10^{10}
4	56%	4.60×10^{10}
5	83%	3.96×10^{10}
6	100%	1.53×10^{10}

*M_i is the applied moment and M_{ult} is the ultimate bending moment (axial load of 50 kips)

The selection of the precise values of EI to use in a particular problem is, of course, a complex matter. The EI of the gross section where E is computed as $57,000(f'_c)^{0.5}$ can be used to get the bending stiffness with reasonable accuracy. The procedure suggested above may be used to improve the computations, especially in computing the deflections.

Results from Computer Solutions

The results from the computer solutions for the drilled shaft are shown in Fig. 46. The bending stiffness that was employed in the solutions was 6.96×10^{10} lb-sq in. Two cases are shown in Fig. 46, one in which the pile head was assumed to be fixed against rotation and the other in which the pile head was unrestrained against rotation (free-head). In the practical case, it would be impossible to fix the pile head completely even though the top of the drilled shaft is extended several feet into the mat.

Figure 46 shows that the ultimate bending moment (6.78×10^6 in-lb) is attained for both cases with a lateral load of 116 kips. The deflections for the two cases at the ultimate bending moment ($P_t = 116$ kips) are quite different: 0.44 in for the fixed-head case and 1.87 in for the free-head case. The example information for computer analysis employed in PMEIX and COM624 for this drilled-shaft design can be found in Appendix 2.

If it were possible to construct the drilled shaft with an intermediate degree of fixity, or with a given value of rotational restraint, the maximum bending moment would be smaller than the values shown in Fig. 46a and the groundline deflection would lie between the two curves shown in Fig. 46b.

With regard to the bending moment as a function of depth, for the fixed-head case the maximum moment is negative and occurs at the groundline. The maximum positive moment is smaller and occurs rather close to the top of the pile (see Fig. 42). For the free-head case, the bending moment is zero at the groundline and the maximum bending moment is positive. As may be understood, an intermediate degree of fixity could result in a situation where the negative and the positive bending moments are equal and smaller than that for either of the end points with regard to fixity.

Finally, it is of interest to note that the magnitude of the load that can be carried by the drilled shaft of the example is considerable. A good argument can be made for the use of vertical piles to sustain lateral loads rather than using batter piles.

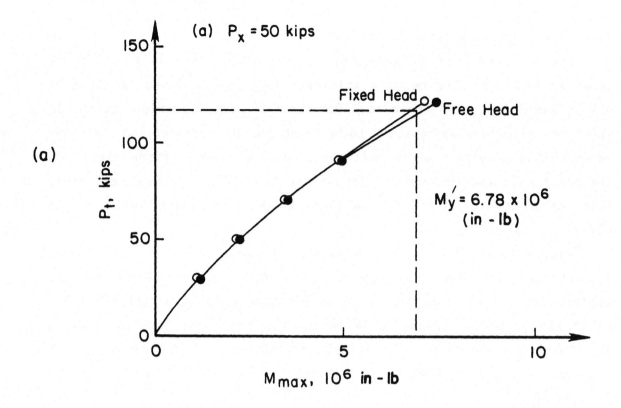
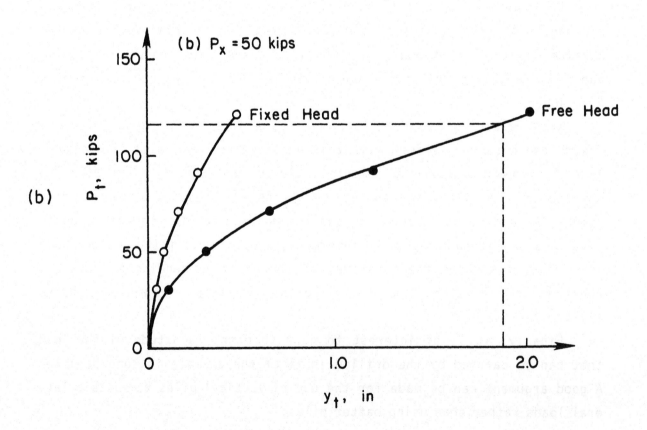

Fig. 46. Results from computer solutions for example problem, drilled shaft supporting bridge abutment

CHAPTER 4. NONDIMENSIONAL SOLUTIONS

4.1 VARIATION OF SOIL MODULUS WITH DEPTH

Prior to presenting the details of nondimensional analysis it is desirable to discuss the nature of the soil modulus. A pile under lateral loading is shown in Fig. 47a and a set of p-y curves is shown in Fig. 47b. As shown in the figure the ultimate value of p and the initial slope of the curves increase with depth, as is to be expected in many practical cases. Also shown in Fig. 47b is the possible deflected shape of the pile under load and the secants to the point on the curves defined by the respective deflection. The soil modulus E_s is defined as p/y and is found by obtaining the slopes of the secants.

The values of E_s so obtained are plotted as a function of depth in Fig. 47c. The line passing through the plotted points defines the variation of E_s with depth. In the case depicted in Fig. 47, the following equation defines the variation in the soil modulus.

$$E_s = kx \qquad (58)$$

It is of interest to note that neither E_s nor k are constants but each of them decrease as the load and deflection increase.

In many cases encountered in practice, the value of E_s would not be zero at the groundline and would not increase linearly with depth, as shown in Fig. 47. However, there are two things that suggest that Eq. 58 will frequently define, at least approximately, the variation of the soil modulus with depth. Firstly, the soil strength and stiffness will usually increase with depth. Secondly, the pile deflection will always be larger at and near the groundline.

Furthermore, experience with nondimensional solutions has shown that it is not necessary to pass a curve precisely through the soil-modulus values, as is done by the computer, to obtain an acceptable solution. This point will be illustrated in the example that is presented later in this chapter.

The derivation of the equations for the nondimensional solutions are not shown here but may be seen in detail elsewhere (Reese and Matlock, 1956; Matlock and Reese, 1961). The following sections present the equations and nondimensional curves for three cases: pile head free to rotate, pile head fixed against rotation, and pile head restrained against

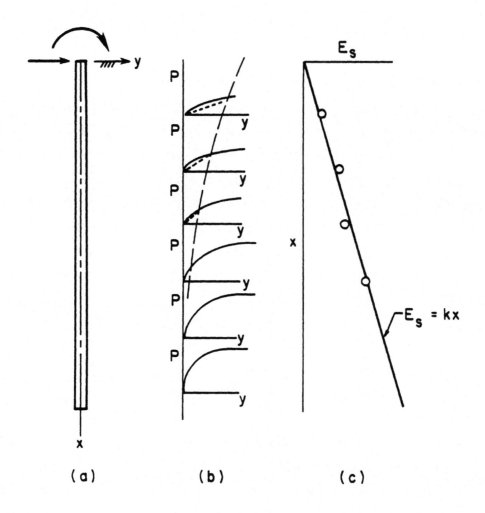

Fig. 47. Form of variation of soil modulus with depth

rotation. The nondimensional solutions are valid only for piles that have constant stiffness EI and with no axial load. These restrictions are not very important in many cases because computer solutions usually show that deflections and bending moments are only moderately influenced by changes in EI and by the presence of an axial load. Also, the principal benefits from the nondimensional method are in checking computer solutions and in allowing an engineer to gain insight into the nature of the problem; thus, precision is not required.

As may be seen by examining published derivations (Matlock and Reese, 1961), nondimensional curves can be developed for virtually any conceivable variation in soil modulus with depth. However, studies show (Reese, 1983) that the utility of some more complex forms of variation, ($E_s = k_1 + k_2 x$, $E_s = k x^n$) is limited when compared to the simpler form ($E_s = kx$).

88

To simplify the use of the nondimensional methods, the procedures in the following sections are presented in step-by-step form.

4.2 PILE HEAD FREE TO ROTATE (CASE I)

The procedure shown in this section may be used when the shear and moment are known at the groundline. A single pile that serves as the foundation for an overhead sign, such as those that cross a highway, is an example of the Case I category. The shear and moment at the groundline may also be known, or computed, for some structural configurations for bridges.

1. Construct p-y curves at various depths by procedures recommended in Chapter 2, with the spacing between p-y curves being closer near the ground surface than near the bottom of the pile.

2. Assume a convenient value of T, perhaps 100 in, the relative stiffness factor. The relative stiffness factor is given as:

$$T = (EI/k)^{1/5} \tag{59}$$

   where

   EI = flexural rigidity of pile,

   k = constant relating the secant modulus of soil reaction to depth ($E_s = kx$).

3. Compute the depth coefficient z_{max}, as follows:

$$z_{max} = \frac{x_{max}}{T} . \tag{60}$$

   where x_{max} equals the embedded length of the pile.

4. Compute the deflection y at each depth along the pile where a p-y curve is available by using the following equation:

$$y = A_y \frac{P_t T^3}{EI} + B_y \frac{M_t T^2}{EI} \tag{61}$$

   where

   A_y = deflection coefficient, found in Fig. 48,

   P_t = shear at top of pile,

   T = relative stiffness factor,

   B_y = deflection coefficient, found in Fig. 49,

   M_t = moment at top of pile, and

   EI = flexural rigidity of pile.

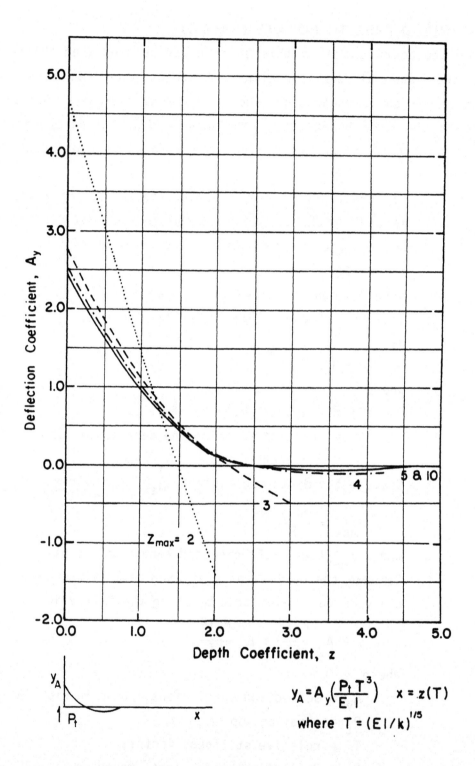

Fig. 48. Pile deflection produced by lateral load at mudline

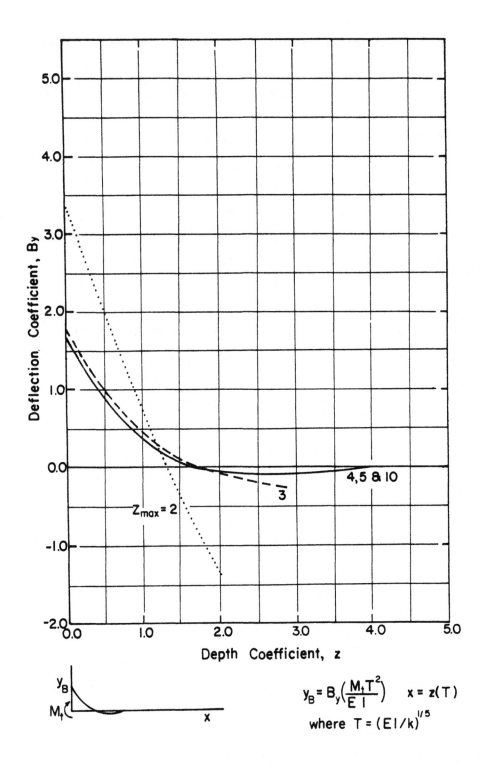

Fig. 49. Pile deflection produced by moment applied at mudline

The particular curves to be employed in getting the A_y and B_y coefficients depend on the value of z_{max} computed in Step 3. The argument for entering Figs. 48 and 49 is the nondimensional depth z, where z is equal to x/T.

5. From a p-y curve, select the value of soil resistance p that corresponds to the pile deflection value y at the depth of the p-y curve. Repeat this procedure for every p-y curve that is available.

6. Compute a secant modulus of soil reaction E_s ($E_s = -p/y$). Plot the E_s values versus depth (see Fig. 47c).

7. From the E_s vs. depth plotted in Step 6, compute the constant k which relates E_s to depth ($k = E_s/x$). Give more weight to the E_s values near the ground surface.

8. Compute a value of the relative stiffness factor T from the value of k found in Step 7. Compare this value of T to the value of T assumed in Step 2. Repeat Steps 2 through 8 using the new value of T each time until the assumed value of T equals the calculated value of T.

9. When the iterative procedure has been completed, the values of deflection along the pile are known from Step 4 of the final iteration. Values of soil reaction may be computed from the basic expression: $p = E_s y$. Values of slope, moment, and shear along the pile can be found by using the following equations:

$$S = A_s \frac{P_t T^2}{EI} + B_s \frac{M_t T}{EI}, \quad (62)$$

$$M = A_m P_t T + B_m M_t, \quad (63)$$

and

$$V = A_v P_t + B_v \frac{M_t}{T}. \quad (64)$$

The appropriate coefficients to be used in the above equations may be obtained from Figs. 50 through 55.

4.3 PILE HEAD FIXED AGAINST ROTATION (CASE II)

The method shown here may be used to obtain a solution for the case where the superstructure translates under load but does not rotate and

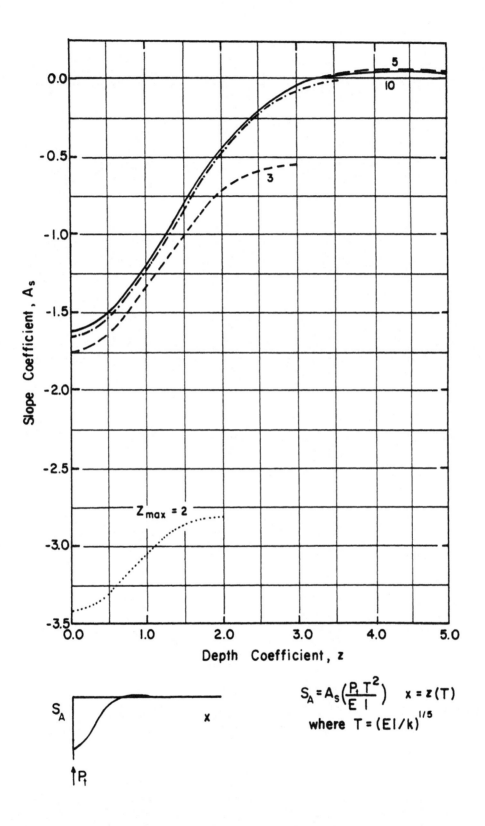

Fig. 50. Slope of pile caused by lateral load at mudline

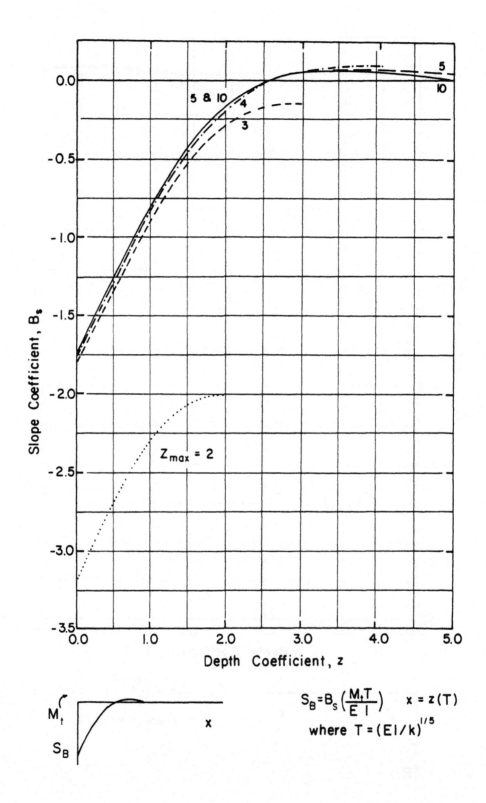

Fig. 51. Slope of pile caused by moment applied at mudline

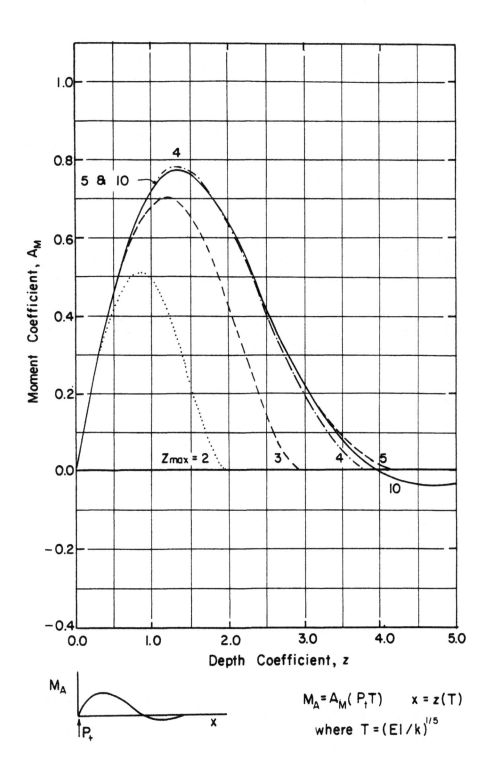

Fig. 52. Bending moment produced by lateral load at mudline

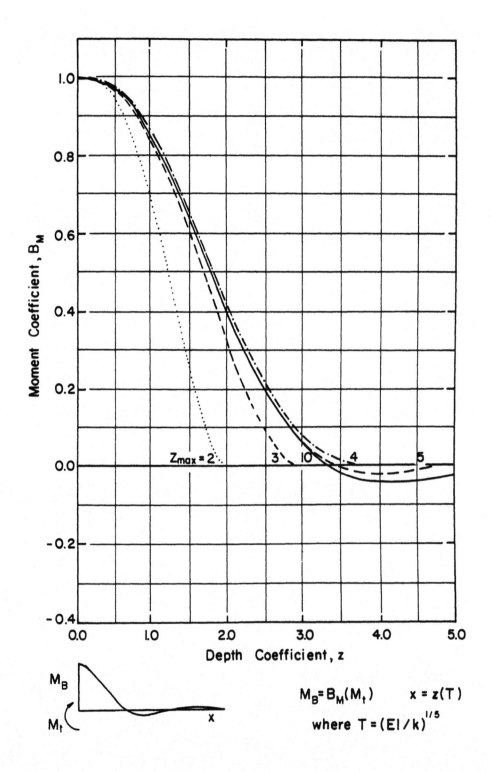

Fig. 53. Bending moment produced by moment applied at mudline

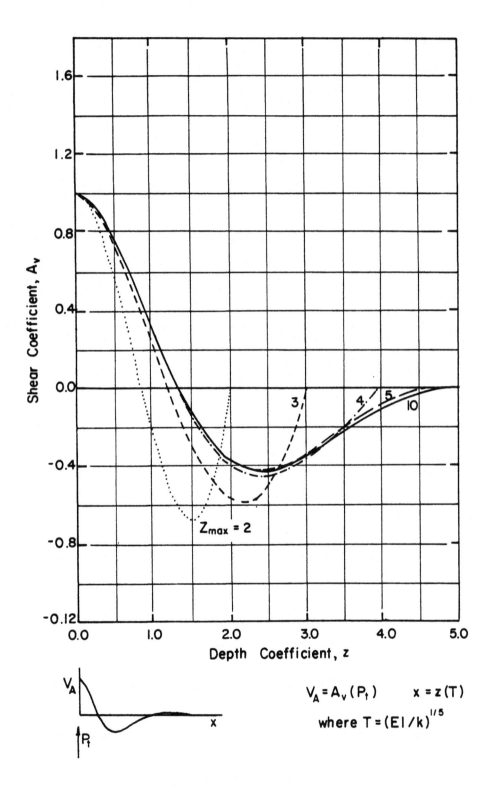

Fig. 54. Shear produced by lateral load at mudline

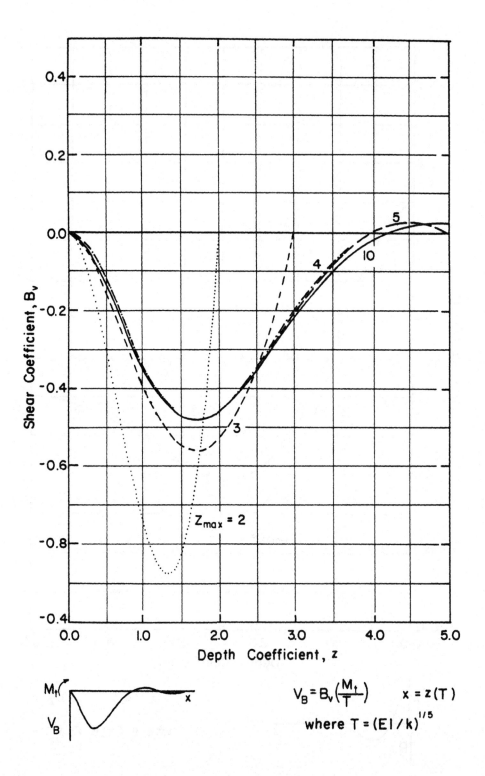

Fig. 55. Shear produced by moment applied at mudline

where the superstructure is very, very stiff in relation to the pile. An example of such a case is where the top of a pile is embedded in a reinforced concrete mat as for a retaining wall or bridge abutment.

1. Perform Steps 1, 2, and 3 of the solution procedure as for free-head piles, Case I.
2. Compute the deflection y_F at each depth along the pile where a p-y curve is available by using the following equation:

$$y_F = F_y \frac{P_t T^3}{EI} . \qquad (65)$$

The deflection coefficients F_y may be found by entering Fig. 56 with the appropriate value of z_{max}

3. The solution proceeds in steps similar to those of Steps 5 through 8 for the free-head case.
4. Compute the moment at the top of the pile M_t from the following equation:

$$M_t = F_{Mt} P_t T. \qquad (66)$$

The value of F_{Mt} may be found by entering Table 12 with the appropriate value of z_{max}.

TABLE 12. MOMENT COEFFICIENTS AT TOP OF PILE FOR FIXED-HEAD CASE

z_{max}	F_{Mt}
2	-1.06
3	-0.97
4	-0.93
5 and above	-0.93

5. Compute values of slope, moment, shear, and soil reaction along the pile by following the procedure in Step 9 for the free-head pile.

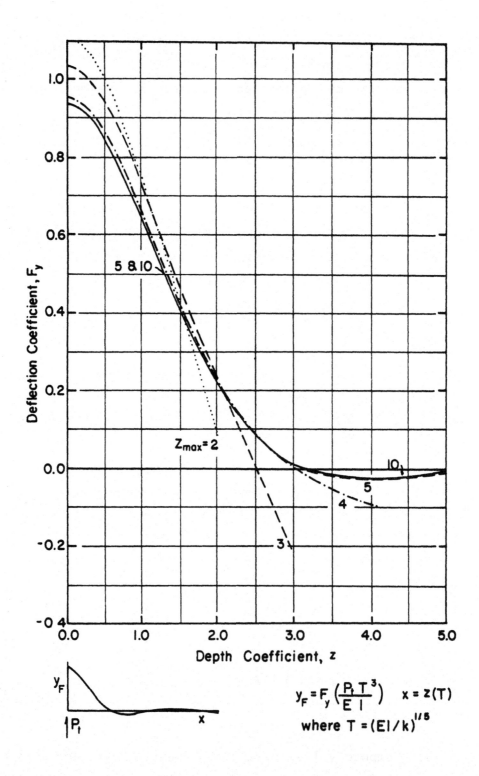

Fig. 56. Deflection of pile fixed against rotation at mudline

4.4 PILE HEAD RESTRAINED AGAINST ROTATION (CASE III)

Case III may be used to obtain a solution for the case where the superstructure translates under load but rotation at the top of the pile is partially restrained. An example of Case III is when the pile is extended and becomes a beam-column of the superstructure. A moment applied to the bottom of the beam-column will result in a rotation, with the moment-rotation relationship being constant. That relationship, then, becomes one of the boundary conditions at the top of the pile.

1. Perform Steps 1, 2, 3 of the solution procedure for free-head piles, Case I.

2. Obtain the value of the spring stiffness k_θ of the pile superstructure system. The spring stiffness is defined as follows:

$$k_\theta = \frac{M_t}{S_t} \quad (67)$$

where

M_t = moment at top of pile, and
S_t = slope at top of pile.

3. Compute the slope at the top of Pile S_t as follows:

$$S_t = A_{st} \frac{P_t T^2}{EI} + B_{st} \frac{M_t T}{EI} \quad (68)$$

where

A_{st} = slope coefficient at z = 0, found in Fig. 50, and

B_{st} = slope coefficient at z = 0, found in Fig. 51.

4. Solve Eqs. 67 and 68 for the moment at the top of the pile M_t.

5. Perform Steps 4 through 9 of the solution procedure for free-head piles, Case I.

This completes the solution of the laterally loaded pile problem for three sets of boundary conditions. The solution gives values of deflection, slope, moment, shear, and soil reaction as a function of depth. To illustrate the solution procedures, an example problem is presented.

4.5 STEP-BY-STEP SOLUTION FOR EXAMPLE PROBLEM 1

The detailed procedure for the solution of a problem by the use of nondimensional methods is presented herein. The first example problem is the same as solved earlier by use of the computer; namely, a steel pile supporting a bridge abutment.

Pile, Soil, and Loading

The data for the example problem are the same as given in Section 3.5, and the normal solution procedure would be as illustrated in that section; however, a hand solution is shown here for one set of loadings. The use of nondimensional solutions serves as a check against the computer solutions. Only in unusual circumstances would the nondimensional method be used to make a complete design.

The loading to be used in the solution is a static lateral load of 36 kips. The nondimensional method does not allow an axial load to be taken into account in solving for the deflection of a pile. The top of the pile is considered to be fixed against rotation and the soil is considered to be a stiff clay above the water table.

p-y Curves

Several p-y curves were tabulated with the output of the computer program shown in Appendix 2. The curves are for depths of 0, 10, 20, 40, 70, 100, and 150 in. The curve for a depth of 10 in will be checked before proceeding with the nondimensional solutions. The procedure given in Section 2.6 will be followed.

Step 1. The following values are tabulated for easy reference: $c = 14$ lb/sq in, $\gamma = 0.069$ lb/cu in, $b = 14.7$ in, $\varepsilon_{50} = 0.007$.

Step 2. Computation of ultimate soil resistance, using Eq. 1 and Eq. 2, and taking the smaller of the two values.

Eq. 1: $p_u = \left[3 + \frac{\gamma}{c} x + \frac{J}{b} x\right] cb$

$= \left[3 + \frac{0.069}{14}(10) + \frac{0.5}{14.7}(10)\right](14)(14.7) = 698$ lb/in.

Eq. 2: $p_u = 9cb = (9)(14)(14.7) = 1852$ lb/in.

Therefore, using the smaller of the two values,

$p_u = 698$ lb/in.

Step 3. Computation of y_{50} from Eq. 3.

Eq. 3: $y_{50} = 2.5\varepsilon_{50}b = (2.5)(0.007)(14.7) = 0.257$ in

Step 4. Computation of points on p-y curve up to $16y_{50}$ (4.116 in). The values of p will be computed for values of y that are shown with computer output and tabulated in Table 13.

Eq. 20 $p/p_u = 0.5(y/y_{50})^{0.25}$ for $0 < y < 16y_{50}$

TABLE 13. VALUES OF p VERSUS y FOR SOIL PILE SYSTEM AT A DEPTH OF 10 INCHES, EXAMPLE PROBLEM 1

y, in	y/y_{50}	$(y/y_{50})^{0.25}$	P, lb/in
0.274	1.067	1.016	355
0.549	2.136	1.209	422
0.823	3.202	1.338	467
1.098	4.272	1.438	502
1.372	5.339	1.520	530
1.646	6.405	1.591	555
1.921	7.475	1.653	577
2.195	8.541	1.710	597
2.470	9.611	1.761	614
2.744	10.677	1.808	631
3.018	11.743	1.851	646
3.293	12.813	1.892	660
3.567	13.879	1.930	674
3.842	14.949	1.966	686
4.116	16.016	2.000	698

Values of y and p presented in the above table agree with the values tabulated in Appendix 2. It is therefore assumed the other p-y values in Appendix 2 are correct. Values for all the p-y curves used in the nondimensional analysis are tabulated in Table 14. The p-y curves for each depth are shown in Fig. 57.

Find Value of Relative Stiffness Factor T

The value of T must be found by using an iterative procedure; that is, computations are continued until the value of T tried is equal to the value obtained. As shown in the following paragraphs, graphical procedures are useful in obtaining the solution.

TABLE 14. COMPUTED p-y CURVES FOR EXAMPLE PROBLEM 1

Deflection, in \ Depth, in	p, lb/in						
	0.	10.	20.	40.	70.	100.	150.
0	0	0	0	0	0	0	0
0.274	314	355	395	477	599	721	925
0.549	373	422	470	567	712	857	1100
0.823	413	467	520	627	788	949	1217
1.098	444	502	559	674	847	1020	1308
1.372	469	530	591	713	895	1078	1383
1.646	491	555	618	746	937	1128	1448
1.921	510	577	643	775	974	1173	1505
2.195	528	597	665	802	1007	1212	1556
2.470	543	614	684	826	1037	1249	1602
2.744	558	631	703	848	1065	1282	1645
3.018	571	646	720	868	1090	1313	1684
3.293	584	660	735	887	1114	1342	1721
3.567	596	674	750	905	1137	1369	1756
3.842	607	686	764	922	1158	1395	1788
4.116	617	698	778	938	1178	1419	1820
5.145	617	698	778	938	1178	1419	1820

<u>Trial 1</u>. Other than experience with similar problems, there is little to guide in the selection of the starting value of T; thus, an approximate value of T is assumed

Try T = 100 in

z_{max} = 600/100 = 6 "Long" pile

$y_F = F_y P_t T^3 / EI$

where

P_t = 36,000 lb

EI = $(29 \times 10^6)(904)$ = 2.6216×10^{10} lb-sq in.

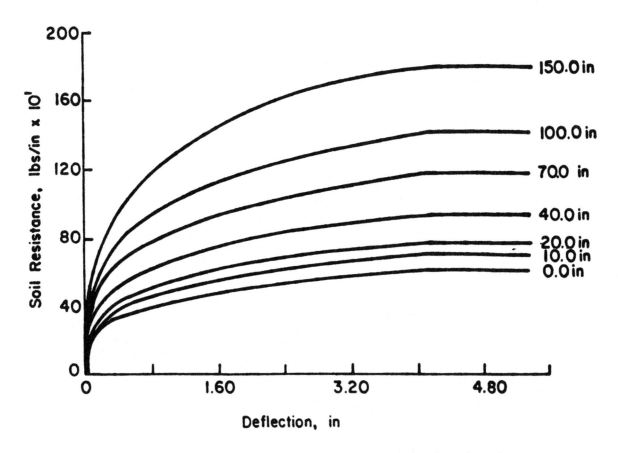

Fig. 57. Computed p-y curves for Example Problem 1

The values of z are computed at depths where p-y curves have been prepared. Equation 65 may be simplified as follows:

$$y_F = 1.4876 \, F_y.$$

Values of F_y may be obtained from Fig. 56 and the y's can be computed. The corresponding values of p may be obtained from Table 14 by interpolation (with slight errors because the table does not precisely represent curves), or values of p can be obtained directly from Eq. 20. The value E_s can then be computed by dividing p by y. Results of these computations are shown in Table 15. The next step is to plot E_s versus x as shown in Fig. 58a and to determine k from the slope of the line fitted as shown in Fig. 58. The value of k was calculated to be 12.0 lb/cu in which corresponds to a value of 73.8 in for T (Eq. 59). Because the plotted points of E_s versus x do not fall on a line defined by $E_s = kx$, an average line was selected by eye.

105

TABLE 15. VALUES OF E_s VERSUS DEPTH FOR TRIAL 1, EXAMPLE PROBLEM 1

x, in	z = x/T	F_y	y_F, in	p, lb in	E_s, lb/sq in
0	0	0.930	1.383	469	340
10	0.1	0.925	1.376	530	385
20	0.2	0.912	1.356	589	434
40	0.4	0.869	1.293	702	543
70	0.7	0.755	1.123	851	758
100	1.0	0.622	0.925	975	1054
150	1.5	0.394	0.586	1115	1904

<u>Trial 2</u>. Because the obtained value of T in Trial 1 was much less than the value tried, a smaller value of T was selected for the second trial.

Try T = 40 in

z_{max} = 600/40 = 15 "Long" pile

y_F = 0.095209 F_y

Using the same procedure outlined in Trial 1, values of E_s were determined and are presented in Table 16.

The values of E_s determined from Trial 2 are plotted versus x in Fig. 58a. The straight line fit through the E_s values is seen to be quite approximate. Experience has shown that better results are obtained if the straight line (E_s = kx) is fitted through the points closer to the groundline rather than to the lower points.

<u>Interpolation for T</u>. The plot in Fig. 58b shows a graphical interpolation for T; as may be seen, a value of 45.6 in was obtained. A final trial could be made as a check by using T = 45.6 in; however, the fitting of E_s = kx to the plotted points is so approximate that other trials would probably not be very productive.

Even though the interpolated value of T is not exact, previous experience has shown that the procedure for obtaining the relative stiffness factor will yield useful results. Therefore, the example will continue by

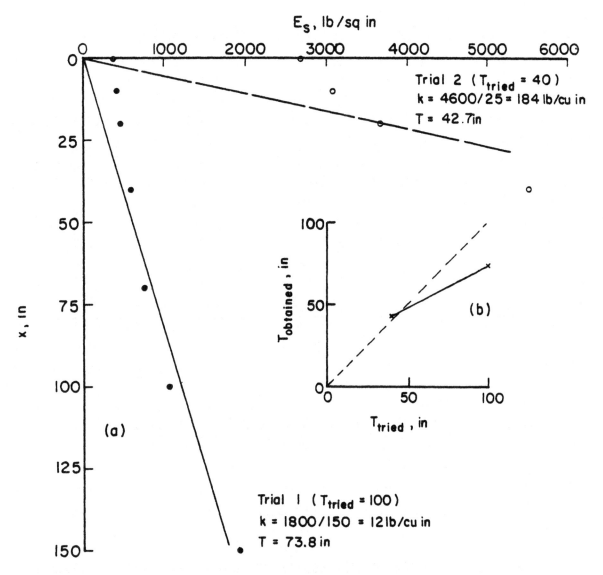

Fig. 58. Graphical solution for relative stiffness factor, Example Problem 1

using the value of T to obtain computed values of deflection and bending moment of the pile as a function of depth.

<u>Computation of Deflection and Bending Moment</u>

Substitution of soil and pile parameters into Eq. 65 produces the following result:

$$y_F = F_y \frac{(39,000)(45.6)^3}{(2.6216 \times 10^{10})} = 0.141 \ F_y.$$

Values of F_y may be obtained from Fig. 56 and the computation of the deflection is shown in Table 17.

TABLE 16. VALUES OF E_s VERSUS DEPTH FOR TRIAL 2, EXAMPLE PROBLEM 1

x, in	z = x/T	F_y	y_F, in	p, lb/in	E_s, lb/sq in
0	0	0.930	0.0885	237	2682
10	0.25	0.905	0.0862	266	3089
20	0.50	0.833	0.0793	289	3653
40	1.00	0.622	0.0592	326	5504
70	1.75	0.295	0.0280	340	12132
100	2.50	0.088	0.0084	303	36086
150	3.75	-0.023	-0.0022	271	123307

TABLE 17. VALUES OF DEFLECTION AND BENDING MOMENT ALONG PILE LENGTH FOR T = 45.6 INCHES, EXAMPLE PROBLEM 1

z	x in	F_y	y_F in	A_m	M_A in-lb	B_m	M_B in-lb	M in-lb
0.0	0	0.930	0.131	0.0	0.0	1.000	-1.654×10^6	-1.654×10^6
0.4	18	0.868	0.122	0.385	0.685×10^6	0.990	-1.637×10^6	-0.953×10^6
0.8	36	0.714	0.101	0.653	1.161×10^6	0.910	-1.505×10^6	-0.344×10^6
1.2	55	0.525	0.074	0.767	1.364×10^6	0.775	-1.282×10^6	0.082×10^6
1.6	73	0.355	0.050	0.754	1.341×10^6	0.593	-0.981×10^6	0.360×10^6
2.0	91	0.205	0.029	0.630	1.120×10^6	0.400	-0.662×10^6	0.459×10^6
2.4	109	0.097	0.014	0.463	0.823×10^6	0.238	-0.394×10^6	0.430×10^6
2.8	128	0.029	0.004	0.300	0.533×10^6	0.109	-0.180×10^6	0.353×10^6
3.2	146	-0.007	-0.001	0.162	0.288×10^6	0.022	-0.036×10^6	0.252×10^6
3.6	164	-0.020	-0.003	0.058	0.103×10^6	-0.025	0.041×10^6	0.144×10^6
4.0	182	-0.020	-0.003	0.0	0.0	-0.040	0.066×10^6	0.066×10^6

The value of the moment at the top of the pile can be computed using Eq. 66, as follows:

$$M_t = (-0.93)(39{,}000)(45.6) = -1.654 \times 10^6 \text{ in-lb}.$$

The moment along the pile can be computed using Eq. 63. Substitution into that equation yields the following:

$$M = A_m(39{,}000)(45.6) + B_m M_t$$
$$= 1.778 \times 10^6 A_m - 1.654 \times 10^6 B_m.$$

The computation of the bending moment is shown in Table 17. Values of A_m and B_m were obtained from Figs. 52 and 53. (Alternately, values of A_m and B_m could be obtained from the table of nondimensional coefficients, Table 21, shown with Example 2.)

Comparison with Computer Solution

The comparison between the nondimensional solution and the computer solution for bending moment versus depth is shown in Fig. 59. A similar comparison for deflection is shown in Fig. 60. As expected, the agreements are not perfect. However, the agreement is close enough to give confidence in the computer solution. As may be noted in the equations for bending moment and deflection, the relative stiffness factor T is raised to the third power in the deflection equation but only to the first power in the bending moment equation; thus, the bending moment curves are less sensitive to variations in the value of T and should generally compare more favorably.

The above computation procedures can be simplified by using some forms which were designed for engineering practice. These forms are included in the appendix. To illustrate the use of these forms, a second example is shown below.

4.6 STEP-BY-STEP SOLUTION FOR EXAMPLE PROBLEM 2

The second example solution by the use of nondimensional methods is presented in the following paragraphs. The example is similar to that of a pile supporting an overhead-sign structure.

Pile, Soil, and Loading

The data for this example problem are given in Fig. 61 and consist of a cyclic lateral load of 10 kips applied 30 ft above the ground surface. The pile is embedded 60 ft in clay with an undrained shear strength of 1000 lb/sq ft.

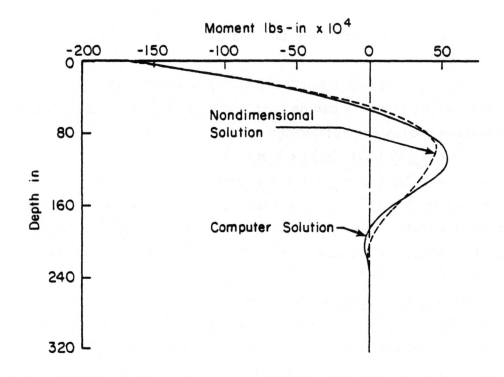

Fig. 59. Comparison of bending moment from computer and from nondimensional solution, Example Problem 1

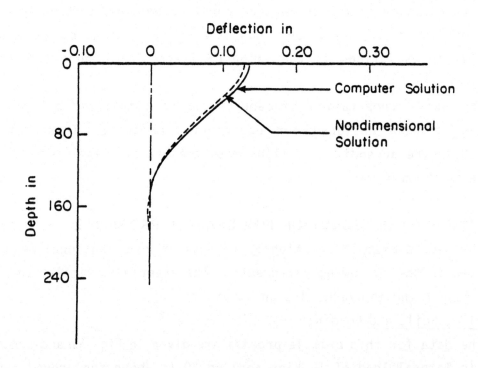

Fig. 60. Comparison of deflection from computer and from nondimensional solution, Example Problem 1

Pile: 24-in O.D. by 0.5 in wall thickness
L = 60 ft = 720 in
EI = 7.39 x 10¹⁰ lb-sq in

Soil: Stiff clay above water table
c = 1000 lb sq/ft = 6.944 lb/sq in
γ = 110 lb/cu ft = 0.0637 lb/cu in
ε_{50} = 0.01

Loading: P_t = 10 kips

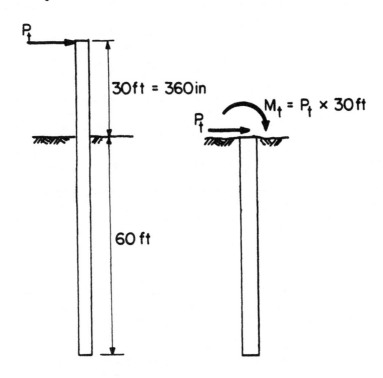

Fig. 61. Pile, soil and loading for Example Problem 2

p-y Curves

Using the computer program, COM624, p-y curves were generated for several depths. Shown in Table 18 are the computed p-y curves for static and cyclic loading at depths of 0, 24, 48, 96, 144, 192 and 288 inches. The verification of these curves is left to the reader.

Find Value of Relative Stiffness Factor T

As demonstrated in the previous example problem, the value of T must be found by iteration. Shown in Tables 19 and 20 are the computations made for Trials 1 and 2. The resulting values of modulus versus depth are

TABLE 18. COMPUTED p-y CURVES FOR EXAMPLE PROBLEM 2

Depth, in		0	24	48	96	144	192	288
y_{static}	y_{cyclic}				p, lb/in			
0.000	0.000	0	0	0	0	0	0	0
0.001	0.003	51	63	75	99	123	147	152
0.015	0.04	100	123	147	195	243	291	299
0.24	0.67	199	247	294	390	485	580	596
0.60	1.68	250	310	370	490	610	730	750
1.24	3.48	300	372	444	588	731	875	899
2.50	7.00	357	443	529	700	872	1043	1072
5.00	14.00	425	527	629	833	1036	1240	1274
9.60	26.88	500	620	740	980	1220	1460	1500

presented in Fig. 62. Values of the nondimensional coefficients, A_y and B_y, were obtained from Table 21. (Alternately, Figs. 48 and 49 could have been used to obtain values of A_y and B_y.)

A final value of 88 in for T was selected based on the results of Trials 1 and 2. The method used to obtain this value of T is illustrated in Fig. 63 and is identical to the method used in the previous example problem.

Computation of Deflection and Bending Moment

Shown in Table 22 are the values of moment and deflection versus depth calculated according to Eqs. 61 and 63. Values of deflection, y, were calculated as follows:

$$y = A_y \frac{P_t T^3}{EI} + B_y \frac{M_t T^2}{EI} \qquad (61)$$

with values of A_y and B_y obtained from Figs. 48 and 49, or from Table 21. Values of bending moment, M, were calculated as follows:

TABLE 19. NONDIMENSIONAL ANALYSIS OF LATERALLY LOADED PILES WITH PILE HEAD FREE TO ROTATE, TRIAL 1 FOR EXAMPLE PROBLEM 2

P_t = __10,000__ lbs M_t = __3.6×10^6__ in-lbs EI = __7.39×10^{10}__ lb-in^2

Trial __1__ $k_{assumed}$ = __7.39__ lb/in^3 (or $T_{assumed}$ = __100__ in)

$T = \left(\dfrac{EI}{k}\right)^{1/5}$ = _____ in $z_{max} = \dfrac{L}{T} = 720/100 = 7.2$

Depth	Depth Coefficient	Deflection Coefficient	Deflection Coefficient	Deflection			Soil Resistance	Soil Modulus
in					in		lb/in	lb/in^2
x	$z = \dfrac{x}{T}$	A_y from Tab. 21	B_y from Tab. 21	$y = A_y \dfrac{P_t T^3}{EI} + B_y \dfrac{M_t T^2}{EI}$			p from p-y curve	$E_s = -\dfrac{p}{y}$
0	0	2.43	1.62	0.33	0.79	1.12	222	198
24	0.24	2.00	1.22	0.27	0.59	0.86	259	301
48	0.48	1.64	0.87	0.22	0.42	0.64	287	448
96	0.96	1.00	0.38	0.14	0.19	0.33	285	864
144	1.44	0.50	0.09	0.06	0.04	0.10	266	2660
192	1.92	0.18	-0.055	0.02	-0.02			
288	2.88	-0.073	-0.10	-0.01	-0.04	-0.05		

$k = \dfrac{E_s}{x} = 9.5$ lb/in^3 $T_{obtained} = \left(\dfrac{EI}{k}\right)^{1/5} = 95.1$ in

TABLE 20. NONDIMENSIONAL ANALYSIS OF LATERALLY LOADED
PILES WITH PILE HEAD FREE TO ROTATE,
TRIAL 2 FOR EXAMPLE PROBLEM 2

P_t = __10,000__ lbs M_t = __3.6 x 10^6__ in-lbs EI = __7.39 x 10^{10}__ lb-in^2

Trial __2__ $k_{assumed}$ = __22.55__ lb/in^3 (or $T_{assumed}$ = __80__ in)

$T = \left(\dfrac{EI}{k}\right)^{1/5}$ = _____ in

$z_{max} = \dfrac{L}{T} = 720/80 = 9$

Depth	Depth Coefficient	Deflection Coefficient	Deflection Coefficient	Deflection			Soil Resistance	Soil Modulus
in				in			lb/in	lb/in^2
x	$z = \dfrac{x}{T}$	A_y from Tab. 21	B_y from Tab. 21	$y = A_y \dfrac{P_t T^3}{EI} + B_y \dfrac{M_t T^2}{EI}$			p from p-y curve	$E_s = -\dfrac{p}{y}$
0	0	2.43	1.62	0.168	0.505	0.67	199	297
24	0.3	1.95	1.14	0.135	0.355	0.49	212	433
48	0.6	1.50	0.75	0.104	0.234	0.34	217	638
96	1.2	0.74	0.22	0.051	0.069	0.12	220	1833
144	1.8	0.25	-0.03	0.017	-0.009	0.01	146	14600
192	2.4		-0.11		-0.034	-0.03		
288	3.6	-0.07	-0.05	-0.005	-0.016	-0.02		

$k = \dfrac{E_s}{x} = 18.4$ lb/in^3 $T_{obtained} = \left(\dfrac{EI}{k}\right)^{1/5} = 83.3$ in

TABLE 21. NONDIMENSIONAL COEFFICIENTS

Nondimensional Coefficients $\qquad$ $E_s = kx$, $z_{max} = 10$

z	A_y	A_S	A_M	A_v	A_p	B_y	B_S	B_M	B_v	B_p
.0	2.435	-1.623	.000	1.000	.000	1.623	-1.750	1.000	.000	.000
.1	2.273	-1.618	.100	.989	-.227	1.453	-1.650	1.000	-.007	-.145
.2	2.112	-1.603	.198	.956	-.422	1.293	-1.550	.999	-.028	-.259
.3	1.952	-1.578	.291	.906	-.586	1.143	-1.450	.994	-.058	-.343
.4	1.796	-1.545	.379	.840	-.718	1.003	-1.351	.987	-.095	-.401
.5	1.644	-1.503	.459	.764	-.822	.873	-1.253	.976	-.137	-.436
.6	1.496	-1.454	.532	.677	-.897	.752	-1.156	.960	-.181	-.451
.7	1.353	-1.397	.595	.585	-.947	.642	-1.061	.939	-.226	-.449
.8	1.216	-1.335	.649	.489	-.973	.540	-.968	.914	-.270	-.432
.9	1.086	-1.268	.693	.392	-.977	.448	-.878	.885	-.312	-.403
1.0	.962	-1.197	.727	.295	-.962	.364	-.792	.852	-.350	-.364
1.2	.738	-1.047	.767	.109	-.885	.223	-.629	.775	-.414	-.268
1.4	.544	-.893	.772	-.056	-.761	.112	-.482	.688	-.456	-.157
1.6	.381	-.741	.746	-.193	-.609	.029	-.354	.594	-.477	-.047
1.8	.247	-.595	.696	-.298	-.445	-.030	-.245	.498	-.476	.054
2.0	.142	-.464	.628	-.371	-.283	-.070	-.155	.404	-.456	.140
2.5	-.020	-.200	.422	-.424	.049	-.105	-.006	.200	-.350	.263
3.0	-.075	-.040	.225	-.349	.226	-.089	.057	.059	-.213	.268
3.5	-.074	.034	.081	-.223	.257	-.057	.065	-.016	-.095	.200
4.0	-.050	.052	.000	-.106	.201	-.028	.049	-.042	.017	.112
4.5	-.026	.042	-.032	-.027	.117	-.009	.028	-.039	.021	.041
5.0	-.009	.025	-.033	.013	.046	.000	.011	-.026	.029	-.002

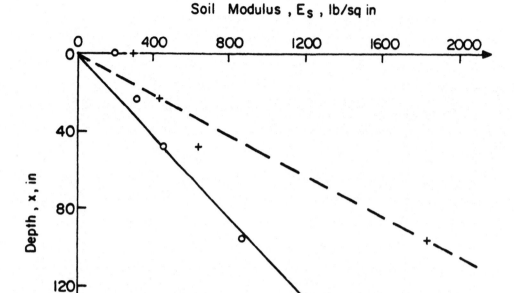

Fig. 62. Trial plots of soil modulus values, Example Problem 2

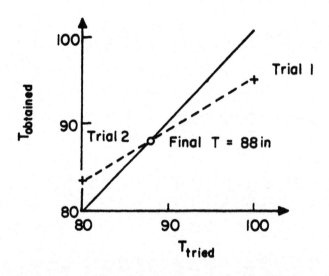

Fig. 63. Interpolation for final values of relative stiffness factor T, Example Problem 2

After substituting, the following equations result:

$$y = A_y P_t T^3/EI + B_y M_t T^2/EI$$
$$= A_y(10,000)(88)^3/(7.39 \times 10^{10})$$
$$+ B_y(3.6 \times 10^6)(88)^2/(7.39 \times 10^{10})$$
$$= 0.0922 A_y + 0.3772 B_y$$
$$M = A_m P_t T + B_m M_t = 8.8 \times 10^5 A_m + 3.6 \times 10^6 B_m$$

TABLE 22. VALUES OF DEFLECTION AND BENDING MOMENT ALONG PILE LENGTH FOR T = 88 INCHES, EXAMPLE PROBLEM 2

z	x	A_y	B_y	y	A_m	B_m	M
	in			in			in-lb
0	0	2.436	1.624	0.837	0	1.000	3.6×10^6
0.25	22	2.032	1.217	0.646	0.245	0.997	3.80
0.50	44	1.644	0.873	0.481	0.458	0.975	3.91
0.75	66	1.284	0.590	0.341	0.621	0.927	3.88
1.00	88	0.963	0.364	0.226	0.726	0.851	3.70
1.25	110	0.686	0.192	0.136	0.770	0.753	3.39
1.50	132	0.458	0.066	0.067	0.761	0.641	2.98
1.75	154	0.277	-0.019	0.018	0.710	0.521	2.50
2.00	176	0.140	-0.072	-0.014	0.628	0.404	2.01
2.25	198	0.042	-0.100	-0.034	0.528	0.296	1.53
2.50	220	-0.023	-0.109	-0.043	0.423	0.202	1.10
2.75	242	-0.061	-0.105	-0.045	0.321	0.124	0.73
3.00	264	-0.080	-0.093	-0.042	0.229	0.064	0.43
3.25	286	-0.083	-0.078	-0.037	0.152	0.022	0.21
3.50	308	-0.078	-0.061	-0.030	0.091	-0.005	0.10
3.75	330	-0.066	-0.044	-0.023	0.048	-0.018	-0.02
4.00	352	-0.052	-0.028	-0.015	0.020	-0.021	-0.06

$$M = A_m P_t T + B_m M_t \tag{63}$$

with values of A_m and B_m obtained from Figs. 52 and 53, or from Table 21.

<u>Comparison with Computer Solution</u>

A comparison between the nondimensional solution and the computer solution for bending moment and deflection versus depth is shown in Fig. 64. As was shown in the previous example problem, the agreement is not perfect; however, the values are reasonably close. Values of moment versus depth from the two analyses agreed well to a depth of about 170 inches. Values of deflection versus depth from the two analyses did not agree as well. Groundline deflections of 0.71 and 0.84 were calculated using the computer program, and the nondimensional analysis, respectively. As dicussed in the previous example problem, a greater discrepancy in deflections is expected since deflections are more sensitive to the parameter T. Another example of nondimensional analysis is shown in Appendix 7.

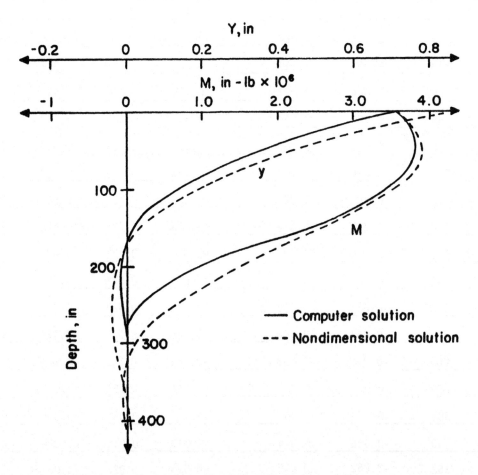

Fig. 64. Comparison of deflection and bending moment from computer and from nondimensional solution, Example Problem 2

CHAPTER 5. BROMS METHOD

The method was presented in three papers published in 1964 and 1965 (Broms, 1964a, 1964b, 1965). Broms developed comprehensive procedures for the design of piles in cohesive and cohesionless soils and his method has been widely used.

5.1 CONCEPTS EMPLOYED IN METHOD

Broms emphasized the computation of the lateral loading at which a pile would fail. Failure was defined in two ways: the development of a plastic hinge or hinges in the pile or, for a short pile, an unlimited deflection. Equations were proposed for computing the ultimate lateral resistance of cohesionless or cohesive soils as a function of depth. The equations assumed an idealized soil profile. The equations of statics were then employed to compute the failure load. The pile head was assumed to be either fixed or free to rotate.

The theory of subgrade reaction, in which there is a linear relationship between load and deflection, was proposed for computing the deflection. It was suggested that the computations of deflection were valid only for a load of up to one-third to one-half of the ultimate load.

5.2 PILES IN COHESIVE SOIL

Assumed Soil Response

Broms adopted distribution of soil resistance shown in Fig. 65. The block shows the value of p_u (force per unit of length) that can be developed as the pile is moved from left to right. The elimination of soil resistance of the top 1.5 diameters of the pile is a result of lower resistance in that zone because a wedge of soil can move up and out when the pile is deflected. The selection of nine times the undrained shear strength times the pile diameter as the ultimate soil resistance, regardless of depth, is based on calculations with the soil flowing from

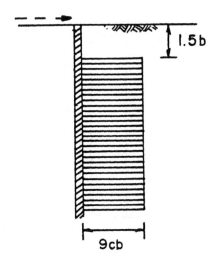

Fig. 65. Assumed distribution of soil resistance for cohesive soil

the front to the back of the pile. Thus, the factor 9cb is consistent with the concept of failure in bearing under lateral load of a section of pile that is deeply embedded.

Free-Head Piles in Cohesive Soil

The problem of "free-head" piles discussed in this section is like the Case I problem (Section 4.2) for nondimensional solutions. As noted in the discussion of that case, an example of the free-head category is the foundation of an overhead sign.

Short, Free-Head Piles in Cohesive Soil.

For piles that are unrestrained against rotation, the patterns that were selected for behavior are shown in Fig. 66. The following equation results from the integration of the upper part of the shear diagram to the point of zero shear (the point of maximum moment)

$$M^{pos}_{max} = P(e + 1.5b + f) - 9cbf^2/2. \tag{69}$$

But the point where shear is zero is

$$f = P/9cb. \tag{70}$$

Therefore,

$$M^{pos}_{max} = P(e + 1.5b + 0.5f). \tag{71}$$

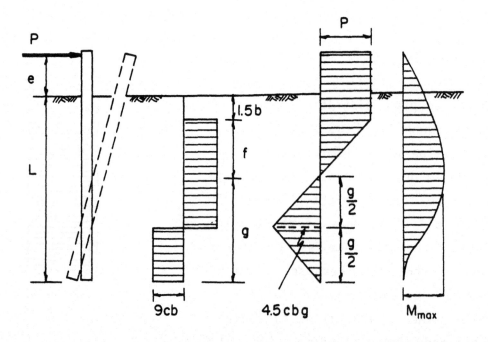

Fig. 66. Deflection, load, shear, and moment diagrams for a short pile in cohesive soil that is unrestrained against rotation

Integration of the lower portion of the shear diagram yields

$$M_{max}^{pos} = 2.25cbg^2. \tag{72}$$

It may be seen that

$$L = (1.5b + f + g). \tag{73}$$

The value of f may be obtained from Eq. 70 and substituted into Eq. 73 to obtain g in terms of P. Equations 71 and 72 are equated and g is substituted to obtain a quadratic equation that will yield the value of P_{ult} that will produce a soil failure. After obtaining a value of P_{ult} the maximum moment can be computed and compared with the moment capacity of the pile. An appropriate factor of safety should be employed.

Broms presented a convenient set of curves for solving the problem of the short-pile (see Fig. 67). The use of the curves eliminates the need to develop and solve a quadratic equation. The use of the curves will be demonstrated later.

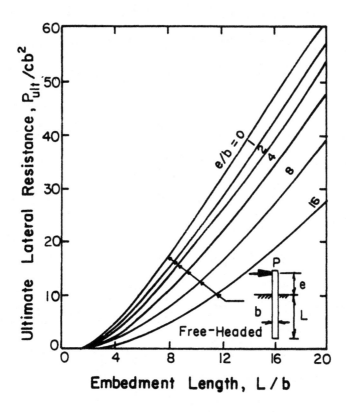

Fig. 67. Design curves for short, free-head piles under lateral load in cohesive soils (after Broms)

Long, Free-Head Piles in Cohesive Soil. As the pile in cohesive soil with the unrestrained head becomes longer, failure will occur with the formation of a plastic hinge at a depth of 1.50b + f. Equation 71 can then be used directly to solve for the ultimate lateral load that can be applied. The shape of the pile under load will be different than that shown in Fig. 66 but the equations of mechanics for the upper portion of the pile remain unchanged.

A plastic hinge will develop when the yield stress of the steel is attained over the entire cross-section.

Broms presented a set of curves for solving the problem of the long pile (see Fig. 68). The use of the curves will be illustrated in the solution of an example problem.

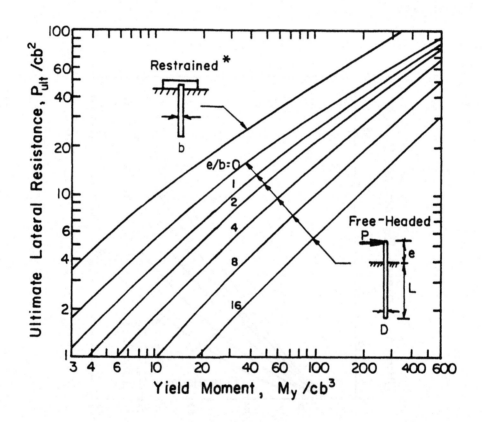

Fig. 68. Design curves for long piles under lateral load in cohesive soils (after Broms)

*Note: The length of the pile for which these curves are valid must be ascertained (see text).

Influence of Pile Length, Free-Head Piles in Cohesive Soil. The preceding sections give procedures for obtaining P_{ult} considering a soil failure (short piles) and a material failure (long piles). Consideration needs to be given to the influence of the pile length. An example problem will be solved where the influence of pile length is investigated.

The following assumptions are made with regard to pile geometry and soil properties.

b = 1 ft(assume 12-in. O.D. steel pipe by 0.75 in wall),

I = 421 in⁴, e = 2 ft, L = 8 ft, and c = 1k/sq ft

f_y = 40 k/sq in (yield moment is 317 ft-k)

The solution begins by assuming that the pile is a short pile. Entering Fig. 67 with L/b of 8 and e/b of 2, a value of 13.5 is obtained for $P_{ult}cb^2$, yielding 13.5 kips for P_{ult}. Alternately, Eqs. 70 through 73 are solved simultaneously and the following quadratic equation is obtained.

$$P^2 + 243P - 3422 = 0$$

and

$$P_{ult} = 13.4 \text{ kips.}$$

Thus, there is good agreement between values obtained from the curves and from the equations. The value of P_{ult} of 13.4 kips is used with Eqs. 70 and 71 with the following result.

$$M_{max} = 13.4(2 + 1.5 + 0.744) = 57 \text{ ft-k}$$

The value of 57 ft-k is far below the value of 317 for a material failure; therefore, a soil failure will occur. The solution would continue by dividing the 13.4 kips by an appropriate factor of safety to obtain a service load.

Consideration is now given to the same problem but with the length increased. If a material failure is assumed, Eqs. 70 and 71 can be used as follows.

$$317 = P_{ult}(2 + 1.5 + \frac{P_{ult}}{18})$$

and

$$P_{ult} = 50.3 \text{ kips.}$$

Alternately, Fig. 68 can be entered with a value of M_y/cb^3 of 317 to obtain a value of P_{ult} of about 50 kips. However, Fig. 68 cannot be read very accurately in some ranges so the equations offer a preferable solution. Again, an appropriate factor of safety should be used to reduce P_{ult} to the service load.

For the example problem, the length at which the short-pile equations cease to be valid may be found by substituting a value of P_{ult} of 50.3 kips into Eq. 70 and solving for f and substituting a value of M_{max} of 317 ft-k into Eq. 72 and solving for g. Equation 73 can then be solved for L. The value of L was found to be 19.0 ft. Thus, for the example problem the value of P_{ult} increases from zero to 50.3 kips as the length of the pile increases from 1.5 ft to 19.0 ft, and above a length of 19.0 ft the value of P_{ult} remains constant at 50.3 kips.

Fixed-Head Piles in Cohesive Soil

The problem of "fixed-head" piles discussed in this section is like the Case II problem (Section 4.3) for nondimensional solutions. As noted in the discussion of that case, an example of the fixed-head category is a pile that has its top embedded in a reinforced concrete mat.

Short, Fixed-Head Piles in Cohesive Soil. For a pile that is fixed against rotation at its top, the mode of failure depends on the length of the pile. For a short pile, failure consists of a horizontal movement of the pile through the soil with the full soil resistance developing over the length of the pile except for the top one and a half pile diameters, where it is expressly eliminated. A simple equation can be written for this mode of failure, based on force equilibrium.

$$P_{ult} = 9cb(L - 1.5b) \tag{74}$$

Broms presented a curve for the short, fixed-head pile in cohesive soil but the curve is omitted because a solution can be obtained so easily with Eq. 74.

Intermediate Length, Fixed-Head Piles in Cohesive Soil. As the pile becomes longer, an intermediate length is reached such that a plastic hinge develops at the top of the pile. Rotation at the top of the pile will occur and a point of zero deflection will exist somewhere along the length of the pile. Figure 69 presents the diagrams of mechanics for the case of the fixed-head pile of intermediate length.

The equation for moment equilibrium for the point where the shear is zero (where the positive moment is maximum) is:

$$M_{max}^{pos} = P(1.5b + f) - f(9cb)(f/2) - M_y.$$

Substituting a value for f from Eq. 70 to simplify,

$$M_{max}^{pos} = P(1.5b + 0.5f) - M_y. \tag{75}$$

Employing the shear diagram for the lower portion of the pile,

$$M_{max}^{pos} = 2.25cbg^2. \tag{76}$$

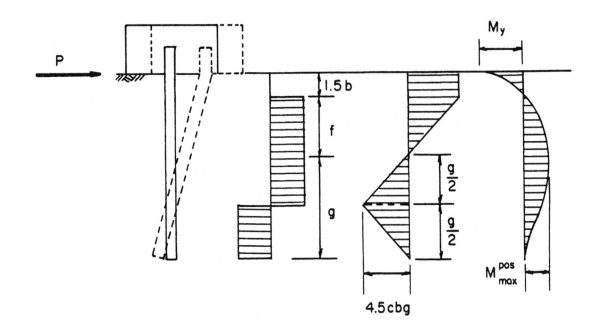

Fig. 69. Deflection, load, shear, and moment diagrams for an intermediate-length pile in cohesive soil that is fixed against rotation at its top

The other equations that are needed to solve for P_{ult} are:

$$L = 1.5b + f + g \tag{77}$$

and

$$f = P/9cb. \tag{78}$$

Equations 75 through 78 can be solved for the behavior of the restrained pile of intermediate length. Broms presented no curves for fixed-head piles of intermediate length in cohesive soil.

<u>Long, Fixed-Head Piles in Cohesive Soil</u>. The mechanics for a long pile that is restrained at its top is similar to that shown in Fig. 69 except that a plastic hinge develops at the point of the maximum positive moment. Thus, the M_{max}^{pos} in Eq. 75 becomes M_y and the following equation results:

$$P = \frac{2M_y}{(1.5b + 0.5f)}. \tag{79}$$

Equations 78 and 79 can be solved to obtain P_{ult} for the long pile. A curve is shown in Fig. 68 for a long pile with fixed head in cohesive soil. However, a note is added to ensure proper use of the curve.

Influence of Pile Length, Fixed-Head Piles in Cohesive Soil. The example problem will be solved for the pile lengths where the pile goes from one mode of behavior to another. All quantities except pile length are the same as used before. Starting with the short pile, an equation can be written for moment equilibrium for the case where the yield moment has developed at the top of the pile and where the moment at its bottom is zero. Referring to Fig. 69, but with the soil resistance only on the right-hand side of the pile, taking moments about the bottom of the pile yields the following equation.

$$PL - 9cb(L - 1.5b)(L - 1.5b)/2 - M_y = 0$$

Summing forces in the horizontal direction yield the next equation.

$$P - 9cb(L - 1.5b) = 0 \quad \text{(same as Eq. 74)}$$

The simultaneous solution of the two equations yield the desired expression.

$$P_{ult} = M_y/(0.5L + 0.75b) \tag{80}$$

Equations 74 and 80 can be solved simultaneously for P_{ult} and L, as follows:

from Eq. 74, $P_{ult} = 9(L - 1.5)$
from Eq. 80, $P_{ult} = 317/(0.5L + 0.75)$
then L = 8.53 ft and P_{ult} = 63.2 k.

For the determination of the length where the behavior changes from that of the pile of intermediate length to that of a long pile, Eqs. 75 through 78 can be used with M_{max}^{pos} set equal to M_y, as follows:

from Eq. 75, $P_{ult} = \dfrac{(2)(317)}{1.5 + 0.5f}$

from Eq. 75, $g = \left(\dfrac{317}{2.25}\right)^{0.5} = 11.87$ ft

from Eq. 77, $L = 1.5 + f + g$
from Eq. 78, $f = P_{ult}/9$
then L = 23.83 ft and P_{ult} = 94.2 k.

In summary, for the example problem the value of P_{ult} increases from zero to 63.2 kips as the length of the pile increases from 1.5 ft to 8.5 ft, increases from 63.2 kips to 94.2 kips as the length increases from 8.5 ft to 23.8 ft, and above a length of 23.8 ft the value of P_{ult} remains constant at 94.2 kips. A factor of safety must be used, of course, to factor the computed loads.

It is of interest to note that the curve in Fig. 68, entered with M_y/cb^3 of 317, yielded a value of P_{ult} of between 90 and 100 kips.

Deflection of Piles in Cohesive Soil

Broms suggested that for cohesive soils the assumption of a coefficient of subgrade reaction that is constant with depth can be used with good results for predicting the lateral deflection at the groundline. He further suggests that the coefficient of subgrade reaction α should be taken as the average over a depth of $0.8\beta L$, where

$$\beta = \left(\frac{\alpha}{4EI}\right)^{0.25} \tag{81}$$

where α = soil modulus (subgrade reaction)

EI = pile stiffness.

Broms presented equations and curves for computing the deflection at the groundline. His presentation for long piles is similar to that shown in the following paragraphs. For short piles and those of intermediate length, the reader may refer to Chapter 4 of <u>Behavior of Piles and Pile Groups Under Lateral Load</u> (Reese, 1983).

For the case where known values of shear and moment are applied at the groundline, the equations for deflection, slope, moment, shear, and soil resistance are shown in Eqs. 82 through 86, respectively.

$$y = \frac{2P_t \beta}{\alpha} C_1 + \frac{M_t}{2EI\beta^2} B_1 \tag{82}$$

$$S = \frac{-2P_t \beta^2}{\alpha} A_1 - \frac{M_t}{EI\beta} C_1 \tag{83}$$

$$M = \frac{P_t}{\beta} D_1 + M_t A_1 \tag{84}$$

$$V = P_t B_1 - 2M_t \beta D_1 \tag{85}$$

$$p = -2P_t \beta C_1 - 2M_t \beta^2 B_1 \tag{86}$$

where

$$A_1 = e^{-\beta x}(\cos \beta x + \sin \beta x) \tag{87}$$
$$B_1 = e^{-\beta x}(\cos \beta x - \sin \beta x) \tag{88}$$
$$C_1 = e^{-\beta x}\cos \beta x \tag{89}$$
$$D_1 = e^{-\beta x}\sin \beta x. \tag{90}$$

Table 23 presents numerical values for A_1, B_1, C_1, and D_1.

TABLE 23. TABLE OF FUNCTIONS FOR PILE OF INFINITE LENGTH

βx	A_1	B_1	C_1	D_1	βx	A_1	B_1	C_1	D_1
0	1.0000	1.0000	1.0000	0.0000	2.4	-0.0056	-0.1282	-0.0669	0.0613
0.1	0.9907	0.8100	0.9003	0.0903	2.6	-0.0254	-0.1019	-0.0636	0.0383
0.2	0.9651	0.6398	0.8024	0.1627	2.8	-0.0369	-0.0777	-0.0573	0.0204
0.3	0.9267	0.4888	0.7077	0.2189	3.2	-0.0431	-0.0383	-0.0407	-0.0024
0.4	0.8784	0.3564	0.6174	0.2610	3.6	-0.0366	-0.0124	-0.0245	-0.0121
0.5	0.8231	0.2415	0.5323	0.2908	4.0	-0.0258	0.0019	-0.0120	-0.0139
0.6	0.7628	0.1431	0.4530	0.3099	4.4	-0.0155	0.0079	-0.0038	-0.0117
0.7	0.6997	0.0599	0.3798	0.3199	4.8	-0.0075	0.0089	0.0007	-0.0082
0.8	0.6354	-0.0093	0.3131	0.3223	5.2	-0.0023	0.0075	0.0026	-0.0049
0.9	0.5712	-0.0657	0.2527	0.3185	5.6	0.0005	0.0052	0.0029	-0.0023
1.0	0.5083	-0.1108	0.1988	0.3096	6.0	0.0017	0.0031	0.0024	-0.0007
1.1	0.4476	-0.1457	0.1510	0.2967	6.4	0.0018	0.0015	0.0017	0.0003
1.2	0.3899	-0.1716	0.1091	0.2807	6.8	0.0015	0.0004	0.0010	0.0006
1.3	0.3355	-0.1897	0.0729	0.2626	7.2	0.0011	-0.00014	0.00045	0.00060
1.4	0.2849	-0.2011	0.0419	0.2430	7.6	0.00061	-0.00036	0.00012	0.00049
1.5	0.2384	-0.2068	0.0158	0.2226	8.0	0.00028	-0.00038	-0.0005	0.00033
1.6	0.1959	-0.2077	-0.0059	0.2018	8.4	0.00007	-0.00031	-0.00012	0.00019
1.7	0.1576	-0.2047	-0.0235	0.1812	8.8	-0.00003	-0.00021	-0.00012	0.00009
1.8	0.1234	-0.1985	-0.0376	0.1610	9.2	-0.00008	-0.00012	-0.00010	0.00002
1.9	0.0932	-0.1899	-0.0484	0.1415	9.6	-0.00008	-0.00005	-0.00007	-0.00001
2.0	0.0667	-0.1794	-0.0563	0.1230	10.0	-0.00006	-0.00001	-0.00004	-0.00002
2.2	0.0244	-0.1548	-0.0652	0.0895					

The equations for a pile whose head is fixed against rotation are given in Eqs. 91 through 95.

$$y = \frac{P_t \beta}{\alpha} A_1 \qquad (91)$$

$$S = \frac{P_t}{2EI\beta^2} D_1 \qquad (92)$$

$$M = -\frac{P_t}{2\beta} B_1 \qquad (93)$$

$$V = P_t C_1 \qquad (94)$$

$$p = -P_t \beta A_1 \qquad (95)$$

For a pile that is partially restrained at its head; that is, where M_t/S_t is known, the solution may be found in Reese (1983).

With regard to values of the coefficient of subgrade reaction, Broms used work of himself and Vesic (1961a, 1961b) for selection of values, depending on the unconfined compressive strength of the soil. The writer believes that the values suggested by Terzaghi (1955) yield results that are compatible with other assumptions and values shown in Table 24 are recommended.

TABLE 24. TERZAGHI'S RECOMMENDATION FOR SOIL MODULUS α FOR LATERALLY LOADED PILES IN STIFF CLAY

Consistency of Clay	Stiff	Very Stiff	Hard
q_u, T/sq ft	1-2	2-4	>4
α, lb/sq in	39-78	78-155	>233

Broms suggested that the use of a constant for the coefficient of subgrade reaction is valid only for a load of one-half to one-third of the ultimate lateral capacity of a pile.

For the example problem, the long pile in cohesive soil that is restrained against rotation at its top will be considered. A value of

P_{ult} of 94.2 kips was computed. A working load of 35 kips is selected for an example computation and, using Table 19, a value of α is selected as 50 lb/sq in. The value of β is

$$\beta = \left(\frac{50}{(4)(30 \times 10^6)(421)}\right)^{0.25} = \frac{1}{178} \text{ in.}$$

The value of βL must be equal to or greater than 4 for the pile to act as a long pile; therefore, the length must be at least 60 feet. The deflection at the top of the pile may be computed from Eq. 91, using a value from Table 24.

$$y_t = \frac{P_t \beta}{\alpha} A_{1t} = \frac{(35,000)(1.0)}{(178)(50)} = 3.9 \text{ in}$$

Effects of Nature of Loading on Piles in Cohesive Soil

The values of soil modulus presented by Terzaghi were apparently for short-term loading. Terzaghi did not discuss dynamic loading or the effects of repeated loading. Also, because Terzaghi's coefficients were for overconsolidated clays only, the effects of sustained loading would probably be minimal. Because the nature of the loading is so important in regard to pile response, some of Broms' remarks are presented here.

Broms suggested that the increase in the deflection of a pile under lateral loading due to consolidation can be assumed to be the same as would take place with time for spread footings and rafts founded on the ground surface or at some distance below the ground surface. Broms suggested that test data for footings on stiff clay indicate that the coefficient of subgrade reaction to be used for long-time lateral deflections should be taken as 1/2 to 1/4 of the initial coefficient of subgrade reaction. The value of the coefficient of subgrade reaction for normally consolidated clay should be 1/4 to 1/6 of the initial value.

Broms suggested that repetitive loads cause a gradual decrease in the shear strength of the soil located in the immediate vicinity of a pile. He stated that unpublished data indicate that repetitive loading can decrease the ultimate lateral resistance of the soil to about one-half its initial value.

Detailed Step-by-Step Procedure for Piles in Cohesive Soil

The procedure that is detailed below is employed in an example problem in the following section.

1. Select the pile for analysis and obtain diameter, length and material properties.

2. Compute the bending stiffness EI and the ultimate bending moment M_y. If the pile geometry varies with depth, the properties for analysis can be selected at the point where the maximum bending moment will occur (at top of pile for fixed-head case and a few diameters from the top for the free-head case).

3. Determine the conditions of restraint at the pile head, whether fixed-head or free-head. If the pile is partially restrained, computations should be done with both cases and judgement employed in estimating the actual behavior.

4. Study the various loadings on the pile and select an appropriate factor of safety for each type of loading. Compute an ultimate load to be used in design.

5. Study the soil profile and select the undrained shear strength c. If the deposit is layered or if the shear strength is not constant with depth, the shear strength should be taken as the average to a depth of five pile diameters below the ground surface.

6. For free-head piles, enter Fig. 67 with a value of L/b and e/b and obtain $P_{ult}cb^2$. Compute P_{ult} for the short pile case. Alternately, Eqs. 70 through 73 may be solved simultaneously for P_{ult}. Use Eqs. 70 and 71 to compute M_{max} and compare it with the value of M_y computed in step 2. If M_{max} is larger than M_y, the equations for the long pile must be used. The value of M_y is employed and Eqs. 70 and 71 are solved for P_{ult}; alternately, P_{ult} may be obtained from Fig. 68. An appropriate factor of safety must be used to factor P_{ult} in order to obtain an allowable service load.

7. For fixed-head piles, the simplest procedure is to use Eqs. 74 and 80 to solve for P_{ult} and L at the point where the pile length changes from the short case to the intermediate case, and to use Eqs. 75 through 78 to solve for P_{ult} and L where the pile length changes from the intermediate case to the long case. The actual length can then be compared with the computed lengths and the following equations can be employed to compute P_{ult}.

   Short piles, Eq. 74

   Intermediate-length piles, Eqs. 75 through 78

   Long piles, Eqs. 78 and 79 (Figure 68 may be used).

An appropriate factor of safety must be used, as indicated previously.

Example Solution for Piles in Cohesive Soil

The example from Section 3.5 is solved using the Broms method. The steps outlined in the previous section are employed.

1. The pile is a 14HP89 with a width (same as diameter in computations) of 14.7in. It is steel with a yield stress of 40 kips/sq in and with a length of 50 ft.
2. The EI is 2.62×10^{10} lb-sq in and its yield moment (ignoring the axial load) is 5840 in-kips (486.7 ft-kips).
3. It is assumed that the pile is fixed against rotation at the groundline.
4. For this example, the ultimate lateral load will be computed. It is assumed that the lateral load is applied at the groundline.
5. The average shear strength c is 2016 lb/sq ft.
6. Solving for the point where the pile goes from a short pile to one of intermediate length, using Eqs. 74 and 80:

    From Eq. 74, $P_{ult} = (9)(2.02)(1.225)[L - 1.5(1.225)]$
    $= 22.27L - 40.92$

    From Eq. 80, $P_{ult} = 486.7/[0.5L + (0.75)(1.225)]$
    $P_{ult}L + 1.8375P_{ult} - 973.4 = 0$

    Solving simultaneously
    $P_{ult} = 111.9$ kips $L = 6.86$ ft

Solving for the point where the pile goes from one of intermediate length to a long pile, using Eqs. 75 through 78:

   From Eq. 75, $P_{ult} = (2)(486.7)/[(1.5)(1.225) + 0.5f]$
   From Eq. 76, $g = [486.7/(2.25)(2.02)(1.225)]^{0.5} = 9.35$ ft
   From Eq. 77, $L = (1.5)(1.225) + f + 9.35 = 11.19 + f$
   From Eq. 78, $P_{ult} = (9)(2.02)(1.225)f = 22.27f$

   Solving for f
   $22.27f = 973.4/(1.8375 + 0.5f)$
   $f = 7.69$ ft
   then $P_{ult} = 171.3$ kips
   and $L = 18.9$ ft.

These computations indicate that the pile length is in the range of long piles and that P_{ult} is 171.3 kips. Equations 78 and 79 will be re-solved to confirm the computations above.

From Eq. 78, $P_{ult} = (9)(2.02)(1.225)f = 22.27f$
From Eq. 79, $P_{ult} = (2)(486.7)/[(1.5)(1.227) + 0.5f]$
equating $40.99f + 11.135f^2 = 973.4$
and $f^2 + 3.68f - 87.42 = 0$ $f = 7.69$ ft (as before)
$P_{ult} = 171.3$ kips.

As a check on the above computations, the restrained-head curve in Fig. 68 is entered with M_y/cb^3 of 131. A value of P_{ult}/cb^2 of about 58 was found; thus, the P_{ult} from the curve was 176 kips, a reasonable check.

At this point, it is of interest to note that the computer solution presented in Chapter 3 indicated an ultimate lateral load of 98 kips, far less than was indicated by the above solution.

Deflection is computed, using the Broms approach, for the service load of 39 kips that was used in the computer solution. From Table 24 a value of α was selected as 78 lb/sq in and β can be then computed.

From Eq. 81
$$\beta = [78/(4)(2.62 \times 10^{10})]^{0.25} = 1/191 \text{ in.}$$

From Eq. 91
$$y_t = \frac{(39,000)(1)}{(78)(191)} = 2.61 \text{ in.}$$

The groundline deflection from the computer solution was 0.14 in. This completes the example using the Broms method for cohesive soil. Another example is shown in Appendix 7.

5.3 PILES IN COHESIONLESS SOIL

Assumed Soil Response

As for the case of cohesive soil, two failure modes are considered; a soil failure and a failure of the pile by the formation of a plastic hinge. With regard to a soil failure in cohesionless soil, Broms assumed that the ultimate lateral resistance is equal to three times the Rankine passive pressure. Thus, at a depth Z below the ground surface the soil resistance per unit of length P_z can be obtained from the following equations.

$$P_z = 3b\gamma Z K_p, \text{ and} \tag{96}$$
$$K_p = \tan^2(45 + \frac{\phi}{2}) \tag{97}$$

where

γ = unit weight of soil
K_p = Rankine coefficient of passive pressure
ϕ = angle of internal friction of soil

Free-Head Piles in Cohesionless Soil

Short, Free-Head Piles in Cohesionless Soil. For short piles that are unrestrained against rotation, a soil failure will occur. The curve showing soil reaction as a function of depth is shaped approximately as shown in Fig. 70. The use of M_a as an applied moment at the top of the pile follows the procedure adopted by Broms. If both P and M_a are acting, the result would be merely to increase the magnitude of e. It is unlikely in practice that M_a alone would be applied.

The patterns that were selected for behavior are shown in Fig. 71. Failure takes place when the pile rotates such that the ultimate soil resistance develops from the ground surface to the center of rotation. The high values of soil resistance that develop at the toe of the pile are replaced by a concentrated load as shown in Fig. 71.

The following equation results after taking moments about the bottom of the pile.

$$P(e + L) + M_a = (3\gamma b L K_p)(\tfrac{1}{2} L)(\tfrac{1}{3} L) \qquad (98)$$

Solving for P when M_a is equal to zero,

$$P = \frac{\gamma b L^3 K_p}{2(e + L)} . \qquad (99)$$

And solving for M_a when P is equal to zero,

$$M_a = 0.5 \gamma b L^3 K_p . \qquad (100)$$

Equations 98 through 100 can be solved for the load or moment, or a combination of the two, that will cause a soil failure. The maximum moment will then be found, at the depth f below the ground surface, and compared with the moment capacity of the pile. The distance f can be computed by solving for the point where the shear is equal to zero.

$$P - (3\gamma b f K_p)(\tfrac{f}{2}) = 0 \qquad (101)$$

Solving Eq. 101 for an expression for f

$$f = 0.816(P/\gamma b K_p)^{0.5} . \qquad (102)$$

The maximum positive bending moment can then be computed by referring to Fig. 71.

$$M_{max}^{pos} = P(e + f) - \frac{K_p \gamma b f^3}{2} + M_a \qquad (103)$$

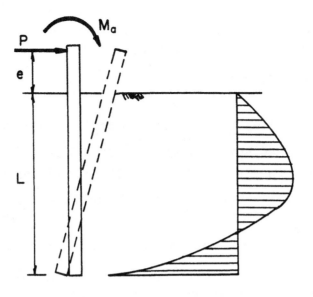

Fig. 70. Failure mode of a short pile in cohesionless soil that is unrestrained against rotation

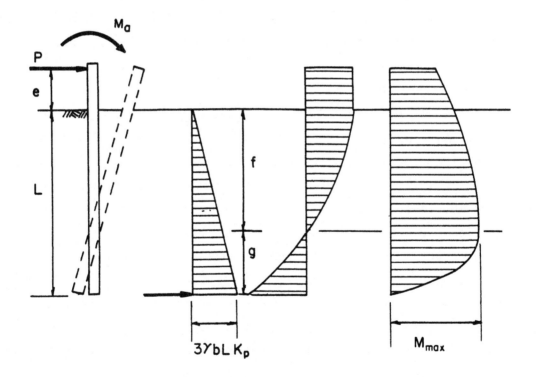

Fig. 71. Deflection, load, shear, and moment diagrams for a short pile in cohesionless soil that is unrestrained against rotation

Or, by substituting expression for Eq. 101 into Eq. 103, the following expression is obtained for the maximum moment.

$$M_{max}^{pos} = P(e + f) - Pf/3 + M_a \qquad (104)$$

An appropriate factor of safety should be used.

   Long, Free-Head Piles in Cohesionless Soil. As the pile in cohesionless soil with the unrestrained head becomes longer, failure will occur with the formation of a plastic hinge in the pile at the depth f below the ground surface. It is assumed that the ultimate soil resistance develops from the ground surface to the point of the plastic hinge. Also, the shear is zero at the point of maximum moment. The value of f can be obtained from Eq. 102. The maximum positive moment can then be computed, and Eq. 104 is obtained. Assuming that M_a is equal to zero, an expression can be developed for P_{ult} as follows:

$$P_{ult} = \frac{M_y}{e + 0.544[P_{ult}/(\gamma b K_p)]^{0.5}} \,. \qquad (105)$$

   Broms presented a set of curves for solving the problem of the long pile in cohesionless soils (see Fig. 72). The logarithmic scales are somewhat difficult to read and it may be desirable to make a solution using Eq. 105. Equations 102 and 104 must be used in any case if a moment is applied at the top of the pile.

   Influence of Pile Length, Free-Head Piles in Cohesionless Soil. There may be a need to solve for the pile length where there is a change in behavior from the short-pile case to the long-pile case. As for the case of the pile in cohesive soils, the yield moment may be used with Eqs. 98 through 103 to solve for the critical length of the pile. Alternatively, a solution could be made with the short-pile equations and the resulting maximum moment could then be compared with the yield moment. If the yield moment is less, the long-pile equations must be used. An example problem is solved where the influence of pile length is investigated. The pile used previously, 12-in O.D., is considered. The angle of internal friction of the sand is assumed to be 34 degrees and the unit weight is assumed to be 55 pounds per cubic foot (the water table is assumed to be above the ground surface). Assume M_a is equal to zero. Equations 97 and 99 yield the following:

$$K_p = \tan^2(45 + \tfrac{34}{2}) = 3.54$$

$$P_{ult} = \frac{(0.055)(1)(8)^3(3.54)}{2(2 + 8)} = 4.98 \text{ kips.}$$

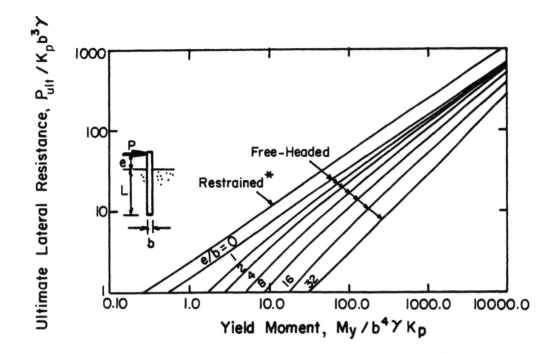

Fig. 72. Design curves for long piles under lateral load in cohesionless soils (after Broms)

*Note: The length of pile for which this curve is valid must be ascertained (see text).

The distance f can be obtained from Eq. 102.

$$f = \left(\frac{4.98}{(1.5)(0.055)(1)(3.54)}\right)^{0.5} = 4.13 \text{ ft}$$

The maximum bending moment can be computed using Eq. 103.

$$M_{max} = (4.98)(2 + 4.13) - \frac{(3.54)(0.055)(1)(4.13)^3}{2} = 23.7 \text{ ft-k}$$

Assuming no axial load, the maximum bending stress f_b is

$f_b = (23.7)(12)(6)/421 = 4.05$ k/sq in.

The computed maximum stress is undoubtedly tolerable, especially when a factor of safety is used to reduce P_{ult}. Broms presented curves for the solution of the case where a short, unrestrained pile undergoes a soil failure; however, Eqs. 98 through 103 are so elementary that such curves are unnecessary.

For the long pile, a value of M_y of 317 in-k is substituted into Eq. 105 with the following result.

$$P_{ult} = \frac{317}{2 + 0.544[P_{ult}/\{(0.055)(1)(3.54)\}]^{0.5}} = 34.36 \text{ kips}$$

Entering the curves in Fig. 70 with a value of $M_y/b^4\gamma K_p$ of 1628, one obtains a value of P_{ult} of about 35 kips.

For the example problem, the value of P_{ult} of 34.36 kips is substituted into Eq. 99 and a value of L of 19.7 ft is computed. Thus, for the pile that is unrestrained against rotation the value of P_{ult} increases from zero when L is zero to a value of 34.36 kips when L is 19.7 ft. For larger values of L, the value of P_{ult} remains constant at 34.36 kips.

Fixed-Head Piles in Cohesionless Soil

Short, Fixed-Head Piles in Cohesionless Soil. For a pile that is fixed against rotation at its top, as for cohesive soils, the mode of failure for a pile in cohesionless soil depends on the length of the pile. For a short pile, the mode of failure will be a horizontal movement of the pile through the soil, with the ultimate soil resistance developing over the full length of the pile. The equation for static equilibrium in the horizontal direction leads to a simple expression.

$$P_{ult} = 1.5\gamma L^2 b K_p \qquad (106)$$

In his presentation Broms presented curves for the short pile that is fixed against rotation. Those curves are omitted here because the equations are so easy to solve.

Intermediate Length, Fixed-Head Piles in Cohesionless Soil. As the pile becomes longer, an intermediate length is reached such that a plastic hinge develops at the top of the pile. Rotation at the top of the pile will occur, and a point of zero deflection will exist somewhere along the length of the pile. The assumed soil resistance will be the same as shown in Fig. 71. Taking moments about the toe of the pile leads to the following equation for the ultimate load.

$$P_{ult} = M_y/L + 0.5\gamma b L^2 K_p \qquad (107)$$

Equation 107 can be solved to obtain P_{ult} for the pile of intermediate length.

Broms presented no curves for piles of intermediate length whose heads are fixed against rotation.

Long, Fixed-Head Piles in Cohesionless Soil. As the length of the pile increases further, the mode of behavior will be that of a long pile. A plastic hinge will form at the top of the pile where there is a negative bending moment, M_y^-, and at some depth f where there is a positive bending moment, M_y^+. The shear at depth f is zero and the ultimate soil resistance is as shown in Fig. 71. The value of f may be determined from Eq.102 but that equation is re-numbered and presented here for convenience.

$$f = 0.816(P/\gamma bK_p)^{0.5} \tag{108}$$

Taking moments at point f leads to the following equation for the ultimate lateral load on a long pile that is fixed against rotation at its top.

$$P_{ult} = \frac{M_y^+ + M_y^-}{e + 0.544[(P_{ult}/\gamma bK_p)]^{0.5}} \tag{109}$$

Equations 108 and 109 can be solved to obtain P_{ult} for the long pile.

Broms presented a curve for the long pile that is fixed against rotation, as shown in Fig. 72. A note is added to ensure the proper use of the curve.

Influence of Pile Length, Fixed-Head Piles in Cohesionless Soil. The example problem will be solved for the pile lengths where the pile goes from one mode of behavior to another. An equation can be written for the case where the yield moment has developed at the top of the short pile. The equation is:

$$P_{ult} = M_y/L + 0.5\gamma bL^2 K_p. \tag{110}$$

Equations 107 and 110 are, of course, identical but the repetition is for clarity. Equations 106 and 110 can be solved for P_{ult} and for L, as follows:

from Eq. 106, $P_{ult} = 0.292L^2$
from Eq. 110, $P_{ult} = 317/L + 0.09735L^2$
then L = 11.77 ft and P_{ult} = 40.4 kips.

For the determination of the length where the behavior changes from that of a pile of intermediate length to that of a long pile, the value of P_{ult} from Eq. 107 may be set equal to that in Eq. 109. It is assumed in this example that the pile has the same yield moment over its entire length.

from Eq. 107, $P_{ult} = 0.09735L^2 + \dfrac{317}{L}$

from Eq. 109, $P_{ult} = \dfrac{634}{2 + 0.544(P_{ult}/0.1947)^{0.5}}$

then L = 20.5 ft and P_{ult} = 56.4 kips

In summary, for the example problem the value of P_{ult} increases from zero to 40.4 kips as the length of the pile increases from zero to 11.77 ft, increases from 40.4 kips to 56.4 kips as the length increases from 11.77 ft to 20.5 ft, and above 20.5 ft the value of P_{ult} remains constant at 56.4 kips.

For the example problem, a value of 68 kips was obtained from Fig. 57 for P_{ult}, which agrees poorly with the computed value. The difficulty probably lies in the inability to read the logarithmic scales accurately.

Deflection of Piles in Cohesionless Soil

Broms noted that Terzaghi (1955) has shown that the coefficient of lateral subgrade reaction for a cohesionless soil can be assumed to increase approximately linearly with depth. As discussed in Chapter 4, for sands, and for many other cases, good results are frequently obtained when using the following equation for the soil modulus.

$$E_s = kx \tag{111}$$

Broms suggested that Terzaghi's values can be used only for computing deflections up to the working load and that the measured deflections are usually smaller than the computed ones except for piles that are placed with the aid of jetting. Terzaghi's recommendations are shown in Table 25. The suggestions for k in Tables 8 and 9 are really not appropriate because those values are meant only to establish the early branch of a p-y curve. Terzaghi's values, on the other hand, were meant to define the deflection curve to a point where a substantial bearing stress has been developed against a pile.

Broms presented equations and curves for use in computing the lateral deflection of a pile; however, the methods presented in Chapter 4 are considered to be appropriate.

As an example problem, the long pile in cohesionless soil that is restrained against rotation at its top is considered. The value of P_{ult} was computed to be 56.4 kips and a working load of 20 kips is selected. Using Table 25, a value of k of 10 lb/cu in is selected. The groundline deflection may be computed from the following equations.

TABLE 25. TERZAGHI'S RECOMMENDATIONS FOR VALUES OF k
FOR LATERALLY LOADED PILES IN SAND

Relative Density of Sand	Loose	Medium	Dense
Dry or moist, k, lb/cu in	3.5 - 10.4	13 - 40	51 - 102
Submerged sand, k, lb/cu in	2.1 - 6.4	8 - 27	32 - 64

$$T = (EI/k)^{0.2} \tag{59}$$

$$y_F = F_y \frac{P_t T^3}{EI} \tag{65}$$

Using Eq. 59, a value of 65.7 in is calculated for T. A value of 0.93 for F_{yt} is obtained from Fig. 56 and the deflection at the groundline is found to be 0.43 in. For the solution to be valid, the length of the pile should be at least 5T or 27.4 ft. Had the pile been shorter, other values of F_{yt} could have been obtained from Fig. 56.

Effects of Nature of Loading on Piles in Cohesionless Soil

Broms noted that piles installed in cohesionless soil will experience the majority of the lateral deflection under the initial application of the load. There will be only a small amount of creep under sustained loads.

Repetitive loading and vibration, on the other hand, can cause significant additional deflection, especially if the relative density of the cohesionless soil is low. Broms noted that work of Prakash (1962) shows that the lateral deflection of a pile group in sand increased to twice the initial deflection after 40 cycles of load. The increase in deflection corresponds to a decrease in the soil modulus to one-third its initial value.

For piles subjected to repeated loading, Broms recommended for cohesionless soils of low relative density that the soil modulus be decreased to one-fourth its initial value and that the value of the soil modulus be decreased to one-half its initial value for soils of high relative density. He suggested that these recommendations be used with caution because of the scarcity of experimental data.

Detailed Step-by-Step Procedure for Piles in Cohesionless Soil

The procedure that is detailed below is employed in an example problem in the following section.

Steps 1 through 4 follow precisely the steps presented in the procedure for cohesive soil.

5. Study the soil profile and select the angle of internal friction ϕ and the unit weight γ. If the deposit is layered or if the relative density of the deposit varies with depth, the soil properties should be taken as the average to a depth of five pile diameters below the ground surface. The submerged unit weight should be used if the soil is below the water table.

6. For free-head piles in cohesionless soil, the simplest procedure is to substitute the value of the ultimate bending moment into Eq. 104 and to solve the resulting equation and Eq. 102 simultaneously to obtain P_{ult}. Equation 98 (or Eq. 99 or Eq. 100) can then be solved for the critical value of L where there is an intersection between the short-pile and long-pile equations.

    For short-pile case: use Eq. 98 to solve for P_{ult} and then use Eqs. 102 and 104 to obtain the maximum bending moment.

    For long-pile case: substitute the value of the ultimate bending moment M_y into Eq. 105 (assuming M_a is zero or that e is modified to reflect M_a) and solve for P_{ult}. Alternatively, Fig. 72 may be entered with $M_y/b^4 \gamma K_p$ and P_{ult} may be found. The logarithmic scales are difficult to read, however.

    An appropriate factor of safety must be used for both the short-pile and long-pile cases.

7. For fixed-head piles in cohesionless soil, Eqs. 106 and 110 can be used to find the P_{ult} and L where the behavior changes from the case of the short-pile to that of the intermediate-length pile. Equations 107 and 109 can be used to find the P_{ult} and L where the behavior changes from the intermediate-length case to the long-pile case. The actual value of L can then be compared to the computed values at the critical points and the case selected to conform to the actual length. The following equations can then be used to compute P_{ult}.

Short piles, Eq. 106
Intermediate-length piles, Eq. 107
Long piles, Eq. 109 (Figure 57 may be used)

As noted previously, an appropriate factor of safety should be used.

Example Solution for Piles in Cohesionless Soil

The example from Section 3.5 is again solved by using the Broms method; however, the soil is changed to sand as shown in step 5 below.

1. The pile is a 14HP89 with a width (b) of 14.7 in. The yield stress is 40 kips/sq in and the length is 50 ft.
2. The EI is 2.62×10^{10} lb-sq in and its yield moment is 5840 in-kips (486.7 ft-kips).
3. The pile is assumed to be fixed against rotation at the groundline.
4. The ultimate lateral load is not given in this example but will be computed. The load is assumed to be applied at the groundline.
5. The angle of internal friction is assumed to be 36° and the unit weight is assumed to be 118 lb/cu ft over a depth of at least five diameters from the groundline.
6. Solving for the point where the pile goes from a short pile to one of intermediate-length, using Eqs. 106 and 110.

From Eq. 106, $P_{ult} = (1.5)(0.118)L^2 (1.225)(\tan^2 63)$
$= 0.835L^2$

From Eq. 110, $P_{ult} = 486.7/L+(0.5)(0.118)(1.225)(3.85)L^2$
$= 486.7/L + 0.278L^2$

Solving simultaneously

$P_{ult} = 76.3$ kips $L = 9.56$ ft.

Solving for the point where the pile goes from one of intermediate length to a long pile, using Eqs. 107 and 109.

From Eq. 107, $P_{ult} = 486.7/L (0.5)(0.118)(1.225)(3.85)L^2$
$= 486.7/L + 0.278L^2$.

From Eq. 109, $P_{ult} = (2)(486.7)/0.544[P_{ult}/(0.118)(1.255)(3.85)]^{0.5}$
$= 1789/(P_{ult}/0.5565)^{0.5}$.

Solving simultaneously

$P_{ult} = 121$ kips $L = 18.5$ ft.

The above computations show that the length of the pile is in the range of long piles and that P_{ult} is 121 kips. Equation 109 was solved just above so there is no point in making further computations; however, Fig. 72 will be employed to obtain a confirmation, at least an approximate check.

The restrained-head curve in Fig. 72 is entered with $M_y/b^4 \gamma K_p$ of 475. A value of $P_{ult}/K_p b^3 \gamma$ of about 147 was obtained; thus, the P_{ult} from the curve was 123 kips, a reasonable check.

At this point, it is of interest to note that the solution with the computer using COM624 (not shown here) gave an ultimate load of 96 kips for the ultimate bending moment of 5840 in-kips.

Deflection is computed, using the Broms approach, for the service load of 38.4 kips (based on factoring the ultimate lateral load from the computer). From Table 20 a value of k of 20 lb/cu in was selected and the relative stiffness factor T can then be computed

$$T = (2.62 \times 10^{10}/20)^{0.2} = 66.6 \text{ in}$$

and $Z_{max} = L/T = 9 > 5$; therefore, the long-pile equations are used. From Eq. 65

$$y_t = \frac{F_{yt} P_t T^3}{EI}$$

$$= \frac{(0.93)(3.84 \cdot 10^4)(66.6)^3}{2.62 \cdot 10^{10}} = 0.402 \text{ in.}$$

The groundline deflection from the computer solution was 0.231 in. This completes the example using the Broms method for cohesionless soil.

5.4. EXAMPLE SOLUTION OF A DRILLED SHAFT

A drilled shaft is designed to support the highway bridge. The section is assumed to be 30 inches in outside diameter and to have 12 No. 8 rebars placed on a 24-in diameter circle. The ultimate strength of the concrete is assumed to be 4000 lb/sq in and the bending stiffness (EI) is 6.96×10^{10} lb-sq in. Compute the ultimate lateral loading on the pile head assuming free-head conditions and the same soil information shown in example 1.

Solution

1. The drilled shaft with 30-in O.D. has 12 No. 8 rebars placed on a 24-in diameter. The total length is 50 ft and the ultimate concrete strength (f_c) is 4000 lb/sq in.

2. The EI is 6.96×10^{10} lb-sq in and its yield moment can be computed from computer program PMEIX which gives the value $M_{ult} = 6.78 \times 10^6$ in-lb (we will assume $M_{ult} = M_y$).
3. It is assumed that the pile head is free to rotate.
4. The ultimate lateral load acted on the pile head at ground surface will be computed.
5. The average shear strength c is 2016 lb/sq ft.
6. Assume the pile is a short pile and compute P_{ult} from Fig. 67 to check this assumption.

$\quad$ L = 50 ft, b = 30 in = 2.5 ft, e = 0
$\quad$ L/b = 50/2.5 = 20
$\quad$ e/b = 0

Estimate the P_{ult}/cb^2 to be about 60 from Fig. 67
$\quad P_{ult}/cb^2 = 60$
$\quad P_{ult} = (2.016)(2.5)^2(60) = 756$ kips

From Eq. 70, $\quad f = P/9cb = \dfrac{756}{(9)(2.016)(2.5)} = 16.67$ ft

From Eq. 71, $\quad M^{pos}_{max} = P[e + 1.5b + (0.5)(16.67)]$
$\qquad = 756[0 + (1.5)(2.5) +$
$\qquad (0.5)(16.67)]$
$\qquad = 9135$ ft-kips
$\qquad = 1.09 \times 10^8$ in-lb.

Since M^{pos}_{max} (1.09×10^8 in-lb) > M_y (6.78×10^6 in-lb), the long pile equations must be used for this drilled shaft.

7. Use the long pile equations to calculate P_{ult}. After we compute the value of M_y/cb^3, we can find the value of P_{ult}/cb^2 from Fig. 68.

$$M_y/cb^3 = \frac{(6.78)(10^6)}{\left(\dfrac{2016}{144}\right)(30)^3} = 17.9$$

Estimate $P_{ult}/cb^2 = 9$ from Fig. 68 and
$\quad P_{ult} = 9(2.016)(2.5)^2 = 113$ kips

The computer result presented in section 3.7 is 120 kips for P_{ult}.

Deflection is computed at the service load 48 kips which is the value of P_{ult} from the computer result divided by safety factor 2.5.

A value of α was selected as 78 lb/sq in from Table 24 and β can be computed:

$$\beta = \left(\frac{\alpha}{4EI}\right)^{0.25} = \left(\frac{78}{(4)(6.96)(10^{10})}\right)^{0.25} = 0.0041$$

From Eq. 82, $\quad y_t = \dfrac{2P_t \beta}{\alpha} C_1 + \dfrac{M_t}{2EI\beta^2} B_1$

$$= \frac{(2)(48 \cdot 10^3)(0.0041)}{78}(1.0) + 0$$

$$= 5.04 \text{ in.}$$

The groundline deflection from the computer result shown in Fig. 46 was only 0.3 in.

CHAPTER 6. ANALYSIS OF PILE GROUPS

6.1 RESPONSE TO LATERAL LOADING OF PILE GROUPS

There are two general problems in the analysis of pile groups: the computation of the loads coming to each pile in the group, and the determination of the efficiency of a group of closely-spaced piles. Each of these problems will be discussed in the following sections.

The methods that are presented are applicable to a pile group that is symmetrical about the line of action of the lateral load. That is, there is no twisting of the pile group so that no pile is subjected to torsion. Therefore, each pile in the group can undergo two translations and a rotation. However, the method that is presented for obtaining the distribution of loading to each pile can be extended to the general case where each pile can undergo three translations and three rotations (Reese, et al., 1970; O'Neill, et al., 1977; Bryant, 1977).

In all of the analyses presented in this section, the assumption is made that the soil does not act against the pile cap. In many instances, of course, the pile cap is cast against the soil. However, it is possible that soil can settle away from the cap and that the piles will sustain the full load. Thus, it is conservative and perhaps logical to assume that the pile cap is ineffective in carrying any load.

If the piles that support a structure are spaced far enough apart that the stress transfer between them is minimal and if the loading is shear only, the methods presented earlier in this work can be employed. Kuthy, et al. (1977) present an excellent treatment of this latter problem.

6.2 WIDELY SPACED PILES

The derivation of the equations presented in this section is based on the assumption that the piles are spaced far enough apart that there is no loss of efficiency; thus, the distribution of stress and deformation from a given pile to other piles in the group need not be considered. However, the method that is derived can be used with a group of closely-spaced piles but another level of iteration will be required.

Model of the Problem

The problem to be solved is shown in Fig. 73. Three piles supporting a pile cap are shown. The piles may be of any size and placed on any batter and may have any penetration below the groundline. The bent may be supported by any number of piles but, as noted earlier, the piles are assumed to be placed far enough apart that each is 100% efficient.

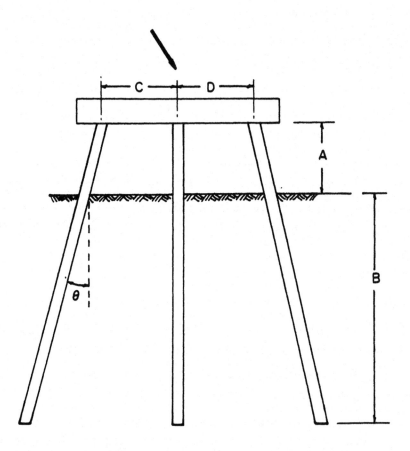

Fig. 73. Typical pile-supported bent

The soil and loading may have any characteristics for which the response of a single pile may be computed.

The derivation of the necessary equations in general form proceeds conveniently from consideration of a simplified structure such as that shown in Fig. 74 (Reese and Matlock, 1966; Reese, 1966). The sign conventions for the loading and for the geometry are shown. A global coordinate system, a-b, is established with reference to the structure. A coordinate system, x-y, is established for each of the piles. For convenience in deriving the equilibrium equations for solution of the problem, the a-b axes are located so that all of the coordinates of the pile heads are positive.

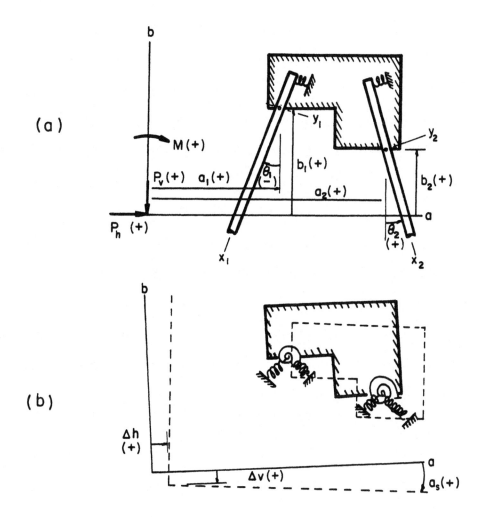

Fig. 74. Simplified structure showing coordinate systems and sign conventions (after Reese and Matlock, 1966)
(a) with piles shown
(b) with piles represented as springs

The soil is not shown, but as shown in Fig. 74b, it is desirable to replace the piles with a set of "springs" (mechanisms) that represent the interaction between the piles and the supporting soil.

If the global coordinate system translates horizontally Δh and vertically Δv and if the coordinate system rotates through the angle α_s, the movement of the head of each of the piles can be readily found. The angle α_s is assumed to be small in the derivation.

The movement of a pile head x_t in the direction of the axis of the pile is

$$x_t = (\Delta h + b\alpha_s) \sin \theta + (\Delta v + a\alpha_s) \cos \theta. \tag{112}$$

149

The movement of a pile head y_t transverse to the direction of the axis of the pile (the lateral deflection) is

$$y_t = (\Delta h + b\alpha_s) \cos \theta - (\Delta v + a\alpha_s) \sin \theta. \tag{113}$$

The assumption is made in deriving Eqs. 112 and 113 that the pile heads have the same relative positions in space before and after loading. However, if the pile heads move relative to each other, an adjustment can be made in Eqs. 112 and 113 and a solution achieved by iteration.

The movements computed by Eqs. 112 and 113 will generate forces and moments at the pile head. The assumption is made that curves can be developed, usually nonlinear, that give the relationship between pile-head movement and pile-head forces. A secant to a curve is obtained at the point of deflection and called the modulus of pile-head resistance. The values of the moduli, so obtained, can then be used, as shown below, to compute the components of movement of the structure. If the values of the moduli that were selected were incorrect, iterations are made until convergence is obtained.

Using sign conventions established for the single pile under lateral loading, the lateral force P_t at the pile head may be defined as follows:

$$P_t = J_y y_t. \tag{114}$$

If there is some rotational restraint at the pile-head, the moment is

$$M_t = -J_m y_t. \tag{115}$$

The moduli J_y and J_m are not single-valued functions of pile-head translation but are functions also of the rotation α_s of the structure. For batter piles, a procedure is given in Appendix 6 for adjusting values of soil resistance to account for the effect of the batter.

If it is assumed that a compressive load causes a positive deflection along the pile axis, the axial force P_x may be defined as follows:

$$P_x = J_x x_t. \tag{116}$$

It is usually assumed that P_x is a single-valued function of x_t.

A curve showing axial load versus deflection may be computed by one of the procedures recommended by several authors (Reese, 1964; Coyle and Reese, 1966; Coyle and Sulaiman, 1967; Kraft, et al., 1981) or the results from a field load test may be used. A typical curve is shown in Fig. 75a.

The computer program or nondimensional methods, presented in Chapters 3 and 4, may be used to obtain curves showing lateral load as a function of lateral deflection and pile-head moment as a function of lateral deflection. The way the pile is attached to the superstructure must be

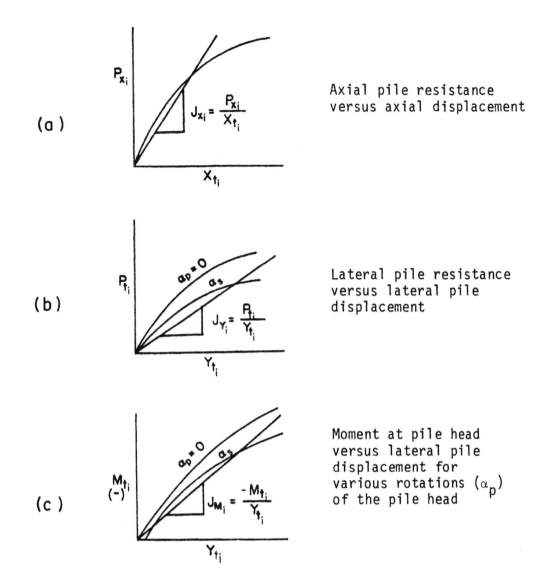

Fig. 75. Set of pile resistance functions for a given pile

taken into account in making the computations. Typical curves are shown in Fig. 75b and 75c.

The forces at the pile head defined in Eqs. 114 through 116 may now be resolved into vertical and horizontal components of force on the structure, as follows:

$$F_v = -(P_x \cos \theta - P_t \sin \theta), \text{ and} \tag{117}$$
$$F_h = -(P_x \sin \theta + P_t \cos \theta). \tag{118}$$

The moment on the structure is

$$M_s = J_m y_t . \tag{119}$$

151

The equilibrium equations can now be written, as follows:

$$P_v + \Sigma F_{v_i} = 0, \qquad (120)$$

$$P_h + \Sigma F_{h_i} = 0, \text{ and} \qquad (121)$$

$$M + \Sigma M_{s_i} + \Sigma a_i F_{v_i} + \Sigma b_i F_{h_i} = 0. \qquad (122)$$

The subscript i refers to values from any "i-th" pile. Using Eqs. 112 through 119, Eqs. 120 through 122 may be written in terms of the structural movements. Equations 123 through 125 are in the final form.

$$\Delta v [\Sigma A_i] + \Delta h [\Sigma B_i] + \alpha_s [\Sigma a_i A_i + \Sigma b_i B_i] = P_v \qquad (123)$$

$$\Delta v [\Sigma B_i] + \Delta h [\Sigma C_i] + \alpha_s [\Sigma a_i B_i + \Sigma b_i C_i] = P_h \qquad (124)$$

$$\Delta v [\Sigma D_i + \Sigma a_i A_i + \Sigma b_i B_i] + \Delta h [\Sigma E_i + \Sigma a_i B_i + \Sigma b_i C_i]$$
$$+ \alpha_s [\Sigma a_i D_i + \Sigma a_i^2 A_i + \Sigma b_i E_i + \Sigma b_i^2 C_i$$
$$+ \Sigma 2 a_i b_i B_i] = M \qquad (125)$$

where

$$A_i = J_{x_i} \cos^2 \theta_i + J_{y_i} \sin^2 \theta_i \qquad (126)$$

$$B_i = (J_{x_i} - J_{y_i}) \sin \theta_i \cos \theta_i \qquad (127)$$

$$C_i = J_{x_i} \sin^2 \theta_i + J_{y_i} \cos^2 \theta_i \qquad (128)$$

$$D_i = J_{m_i} \sin \theta_i \qquad (129)$$

$$E_i = -J_{m_i} \cos \theta_i \qquad (130)$$

The above equations are not as complex as they appear. For example, the origin of the coordinate system can usually be selected so that all of the b-values are zero. For vertical piles, the sine terms are zero and the cosine terms are unity. For small deflections, the J-values can all be taken as constants. Therefore, under a number of circumstances it is possible to solve the above equations by hand. However, if the deflections of the group is such that the nonlinear portion of the curves in Fig. 75 is reached, the use of a computer solution is advantageous. Such a program is available through the Geotechnical Engineering Center, The University of Texas at Austin (Awoshika and Reese, 1971; Lam, 1982).

Detailed Step-by-Step Solution Procedure

1. Study the foundation to be analyzed and select a two-dimensional bent where the behavior is representative of the entire system.

2. Prepare a sketch such that the lateral loading comes from the left. Show all pertinent dimensions.

3. Select a coordinate center and find the horizontal component, the vertical component, and the moment through and about that point.
4. Compute by some procedure a curve showing axial load versus axial deflection for each pile in the group; or, preferably, use the results from a field load test.
5. Use appropriate procedures and compute curves showing lateral load as a function of lateral deflection and moment as a function of lateral deflection, taking into account the effect of structural rotation on the boundary conditions at each pile head.
6. Estimate trial values of J_x, J_y, and J_m for each pile in the structure.
7. Solve Equations 123 through 125 for values of Δv, Δh, and α_s.
8. Compute pile-head movements and obtain new values of J_x, J_y, and J_m for each pile.
9. Solve Equations 123 through 125 again for new values of Δv, Δh, and α_s.
10. Continue iteration until the computed values of the structural movements agree, within a given tolerance, with the values from the previous computation.
11. Compute the stresses along the length of each pile using the loads and moments at each pile head.

Example Problem

Figure 76 shows a pile-supported retaining wall with the piles spaced 8 ft apart. The piles are 14-inches in outside diameter with 4 No. 7 reinforcing-steel bars spaced equally. The centers of the bars are on an 8-in circle. The yield strength of the reinforcing steel is 60 kips/sq in and the compressive strength of the concrete is 2.67 kips/sq in. The length of the piles is 40 ft.

The backfill is a free-draining, granular soil without any fine particles. The surface of the backfill is treated to facilitate a runoff, and weep holes are provided so that water will not collect behind the wall.

The forces P_1, P_2, P_s, and P_w (shown in Fig. 76) were computed as follows: 21.4, 4.6, 18.4, and 22.5 kips, respectively. The resolution of the loads at the origin of the global coordinate system resulted in the

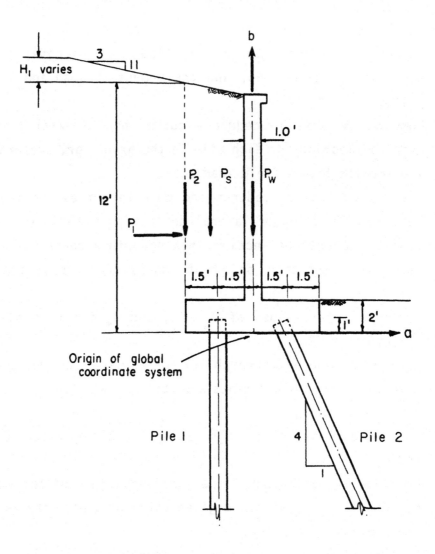

Fig. 76. Sketch of a pile-supported retaining wall

following service loads: P_v = 46 kips, P_h = 21 kips, and M = 40 ft-kips (some rounding was done).

The moment of inertia of the gross section of the pile was used in the analysis. The bending stiffness EI of the piles was computed to be 5.56 x 10^9 lb-sq in. Computer Program PMEIX was run and an interaction diagram for the pile was obtained. That diagram is shown in Fig. 77.

A field load test was performed at the site and the ultimate axial capacity of a pile was found to be 176 kips. An analysis was made to develop a curve showing axial load versus settlement. The curve is shown in Fig. 78.

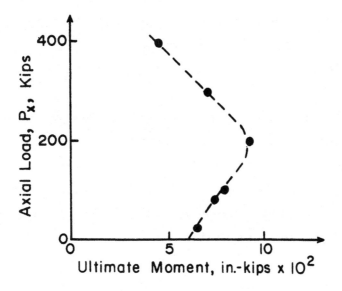

Fig. 77. Interaction diagram of reinforced concrete pile

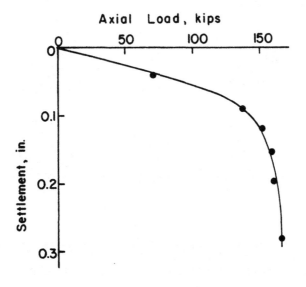

Fig. 78. Axial load versus settlement for reinforced concrete pile

The subsurface soils at the site consist of silty clay. The water content averaged 20% in the top 10 ft and averaged 44% below 10 ft. The water table was reported to be at a depth of 10 ft. There was a considerable range in the undrained shear strength of the clay and an average value of 3 kips/sq ft was used in the analysis. A value of the submerged unit weight of 46 lb/cu ft as employed and the value of ε_{50} was estimated to be 0.005.

In making the computations, the assumption was made that all of the load was carried by the piles with none of the load taken by passive earth pressure or by the base of the footing. It was further assumed that the pile heads were free to rotate. As noted earlier, the factor of safety must be in the loading. Therefore, the loadings shown in Table 26 were used in the preliminary computations.

TABLE 26. VALUES OF LOADING EMPLOYED IN ANALYSES

Case	Loads, kips		moment, ft-kips	Comment
	P_v	P_h		
1	46	21	40	service load
2	69	31.5	60	1.5 times service load
3	92	42	80	2 times service load
4	115	52.5	100	2.5 times service load

Table 27 shows the movements of the origin of the global coordinate system when Eqs. 123 through 125 were solved simultaneously. The loadings were such that the pile response was almost linear so only a small number of iterations were required to achieve convergence. The computed pile-head movements, loads, and moments are shown in Table 28.

TABLE 27. COMPUTED MOVEMENTS OF ORIGIN OF GLOBAL COORDINATE SYSTEM

Case	Vertical movement Δv	Horizontal movement Δh	Rotation α
	in	in	rad
1	0.004	0.08	9×10^{-5}
2	0.005	0.12	1.4×10^{-4}
3	0.008	0.16	1.6×10^{-4}
4	0.012	0.203	8.4×10^{-5}

TABLE 28. COMPUTED MOVEMENTS AND LOADS AT PILE HEADS

Case	Pile 1					Pile 2				
	x_t	y_t	P_x	P_t	M_{max}	x_t	y_t	P_x	P_t	M_{max}
	in	in	kips	kips	in-kips	in	in	kips	kips	in-kips
1	0.005	0.08	9.7	6.0	148	0.02	0.077	38.9	5.8	143
2	0.008	0.12	14.5	9.0	222	0.03	0.116	58.3	8.6	215
3	0.011	0.162	19.3	12.1	298	0.04	0.156	77.7	11.5	288
4	0.013	0.203	24.2	15.2	373	0.06	0.194	97.2	14.3	360

The computed loading on the piles is shown in Fig. 79 for Case 4. The following check is made to see that the equilibrium equations are satisfied.

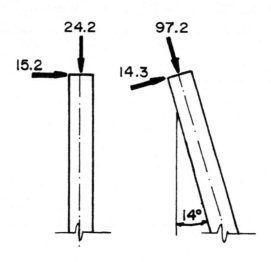

Fig. 79. Pile loading, Case 4

$$\Sigma F_v = 24.2 + 97.2 \cos 14 - 14.3 \sin 14$$
$$= 24.2 + 94.3 - 3.5 = 115.0 \text{ kips OK}$$
$$\Sigma F_h = 15.2 + 14.3 \cos 14 + 97.2 \sin 14$$
$$= 15.2 + 13.9 + 23.6 = 52.7 \text{ kips OK}$$
$$\Sigma M = -(24.2)(1.5) + (97.2 \cos 14)(1.5)$$
$$- (14.3 \sin 14)(1.5)$$
$$= -36.3 + 141.4 - 5.2 = 99.9 \text{ ft-kips OK}$$

Thus, the retaining wall is in equilibrium. A further check can be made to see that the conditions of compatibility are satisfied. One check can be made at once. Referring to Fig. 78, an axial load of 97.2 kips results in an axial deflection of about 0.054 in, a value in reasonable agreement with the value in Table 28. Further checks on compatibility can be made by using the pile-head loadings and Computer Program COM622 to see if the computed deflections under lateral load are consistent with the values tabulated in Table 28.

No firm conclusions can be made concerning the adequacy of the particular design without further study. If the assumptions made in performing the analyses are appropriate, the results of the analyses show the foundation to be capable of supporting the load. As a matter of fact, the piles could probably support a wall of greater height.

6.3 CLOSELY-SPACED PILES

O'Neill (1983) characterizes the problem of closely-spaced piles in a group as one of pile-soil-pile interaction and lists a number of procedures that may be used in predicting the behavior of such groups. He states that none of the procedures should be expected to provide generally accurate predictions of the distribution of loads to piles in a group because none of the models accounts for installation effects. He concludes that there exists a need for more experimental data.

The theory of elasticity has been employed to take into account the effect of a single pile on others in the group. Solutions have been developed (Poulos, 1971b; Banerjee and Davies, 1979) that assume a linear response of the pile-soil system. While such methods are instructive, there is ample evidence to show that soils cannot generally be characterized as linear, homogeneous, elastic materials.

Bogard and Matlock (1983) present a method in which the p-y curve for a single pile is modified to take into account the group effect. Excellent agreement was obtained between their computed results and results from field experiments (Matlock, et al., 1980).

Two approaches to the analysis of a group of closely-spaced piles under lateral load are given in the following paragraphs. One method is closely akin to the use of efficiency formulas and the other method is based on the assumption that the soil within the pile group moves laterally the same amount as do the piles.

Efficiency Formulas

Pile groups under axial load are sometimes designed by use of efficiency formulas. Such a formula is shown as Eq. 131.

$$(Q_{ult})_G = E \cdot n \cdot (Q_{ult})_p \tag{131}$$

where

$(Q_{ult})_G$ = ultimate axial capacity of the group
E = efficiency factor (1 or < 1)
n = number of piles in the group
$(Q_{ult})_p$ = ultimate axial capacity of an individual pile.

Various proposals have been made about obtaining the value of E; for example, McClelland (1972) suggested that the value of E should be 1.0 for pile groups in cohesive soil with center-to-center spacing of eight diameters or more and that E should decrease linearly to 0.7 at a spacing of 3 diameters. McClelland based his recommendations on results from experiments in the field and in the laboratory. It is of interest to note that no differentiation is made between piles that are spaced front to back or side by side or spaced as some other angle between each other.

Unfortunately, as noted earlier, experimental data are limited on the behavior of pile groups under lateral load. Furthermore, the mechanics of the behavior of a group of laterally loaded piles are more complex than for a group of axially loaded piles. Thus, few recommendations have been made for efficiency formulas for laterally loaded groups.

Two different recommendations have been made regarding the modification of the coefficient of subgrade reaction. The Canadian Foundation Engineering Manual (1978) recommends that the coefficient of subgrade reaction for pile groups be equal to that of a single pile if the spacing of the piles in the group is eight diameters. For spacings smaller than eight diameters, the following ratios of the single-pile subgrade reaction were recommended: six diameters, 0.70; four diameters, 0.40; and three diameters, 0.25.

The Japanese Road Association (1976) is less conservative. It is suggested that a slight reduction in the coefficient of horizontal subgrade reaction has no serious effect with regard to bending stress and that the use of a factor of safety should be sufficient in design except in the case where the piles get quite close together. When piles are closer together than 2.5 diameters, the following equation is suggested for computing a factor μ to multiply the coefficient of subgrade reaction for the single pile.

$$\mu = 1 - 0.2(2.5 - L/D) \quad L < 2.5D \tag{132}$$

where

L = center-to-center distance between piles, and

D = pile diameter.

Single-Pile Method

The single-pile method of analysis is based on the assumption that the soil contained between the piles moves with the group. Thus, the pile group with the contained soil can be treated as a single pile of large diameter.

A step-by-step procedure for the use of the method is shown in the following statements.

1. The group to be analyzed is selected and a plan view of the piles at the groundline is prepared.
2. The minimum length is found for a line that encloses the group. If a nine-pile (3 by 3) group consists of piles that are one foot square and three widths on center, the length of the line will be 28 ft.
3. The length found in step 2 is considered to be the circumference of a pile of large diameter; thus, the length is divided by π to obtain the diameter of the imaginary pile.
4. The next step is to determine the stiffness of the group. For a lateral load passing through the tops of the piles, the stiffness of the group is taken as the sum of the stiffness of the individual piles. Thus, it is assumed that the deflection at the pile top is the same for each pile in the group and, further, that the deflected shape of each pile is identical. Some judgement must be used if the piles in the group have different lengths.
5. Then, an analysis is made for the imaginary pile, taking into account the nature of the loading and the boundary conditions at the pile head. The shear and moment for the imaginary large-sized pile is shared by the individual piles according to the ratio of the lateral stiffness of the individual pile to that of the group.

The shear, moment, pile-head deflection, and pile-head rotation yield a unique solution for each pile in the group. As a final step, it is necessary to compare the single-pile solution to that of the group. It could possibly occur that the piles in the group could have an efficiency greater than one, in which case the single-pile solutions would control.

A sketch of an example problem is shown in Fig. 80. It is assumed that steel piles are embedded in a reinforced concrete mat in such a way that the pile heads do not rotate. The piles are 14HP89 by 40 ft long and placed so that bending is about the strong axis. The moment of inertia is 904 in^4 and the modulus of elasticity of 30 x 10^6 lb/sq in. The width of the section is 14.7 in and the depth is 13.83 inches.

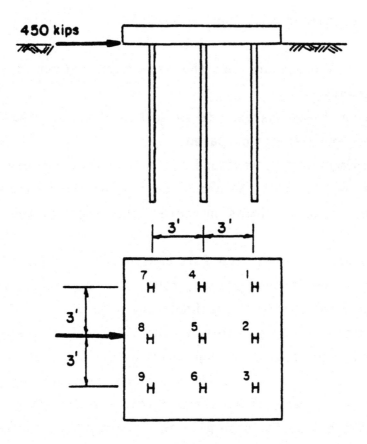

Fig. 80. Plan and elevation of foundation analyzed in example problem

The soil is assumed to be a sand with an angle of internal friction of 34 degrees, and the unit weight is 114 lb/cu ft.

The computer program was run with a pile diameter of 109.4 in and a moment of inertia of 8136 in⁴ (9 times 904). The results were as follows:

y_t = 0.885 in

$M_t = M_{max}$ = 3.60 x 10⁷ in-lb for group
       = 3.78 x 10⁶ in-lb for single pile

Bending stress = 25.3 kips/sq in.

The deflection and stress are for a single pile.

If a single pile is analyzed with a load of 50 kips, the groundline deflection was 0.355 in and the bending stress was 23.1 kips/sq in. Therefore, the solution with the imaginary large-diameter single pile was more critical.

CHAPTER 7. FIELD TESTS OF PILES

7.1 INTRODUCTION

The testing of piles in the field under axial load is a well-established practice and has been common since piles were first used. The testing of piles under lateral loading is less frequent, perhaps because the means of establishing failure of a pile under lateral load has not become common knowledge. However, with the availability of the technology presented in this handbook, there are definite benefits to be gained from the performance of full-scale, field tests of piles under lateral loading. The photograph in Fig. 81 is of a test of two drilled shafts at the Skyway Bridge, Tampa Bay, Florida. The testing arrangement is as described in Fig. 83a.

Fig. 81. Simultaneous testing of two drilled shafts under lateral loading, Skyway Bridge, Tampa Bay, Florida

There are two general reasons for performing field tests of piles: to prove a particular design, and to gain information to allow for a re-design (Reese, 1978). These reasons are valid for the test of a pile under lateral loading and, in addition, in some cases a contribution to technical information can be obtained about the response of a particular soil.

With regard to a proof test of a pile under lateral load, the procedure is not straightforward as with an axial-load test. The response of the pile under lateral load is affected strongly by the way the pile is connected to the superstructure. A single-pile foundation for an overhead sign, for example, will be subjected to a shear and a moment. To try to simulate exactly the pile-head conditions for the sign structure and for a particular kind of loading is impractical if not impossible; therefore, analytical procedures must be employed to interpret the results of virtually any lateral-load test.

Later in this chapter a standard test is described where no internal instrumentation is used in the pile and where only a relatively small amount of instrumentation is used above the groundline. The standard test can be employed to prove any given design and, in some instances, the standard test can provide information for re-design.

Some information is given on a comprehensive testing program where a pile is instrumented internally for the measurement of bending moment along its length. Precise information on soil response at a particular site can be gained from such a testing program, design information will be specific and valuable, and a contribution to the technical literature can be made. The comprehensive program would be expensive and advisable only in special circumstances.

7.2. SELECTION OF TEST SITE

The site selection is simplified if a test is to be performed in connection with the design of a particular structure. However, even in such a case, care should be taken in selecting the precise location of the test pile. In general, the test location should be where the soil profile reveals the weakest condition. In evaluating a soil profile, the soils from the ground surface to a depth equal to five to ten pile diameters are of principal importance. If designed on the basis of the results from the weakest soil, the piles at other places on the construction site should behave more favorably than the test pile.

The selection of a site where a fully instrumented pile is to be tested is usually difficult. The principal aim of such a test is to obtain experimental p-y curves that can be employed in developing predictions of soil response. Thus, the soil at the site must be relatively homogeneous and representative of a soil type for which predictive equations are needed. For many of the past experiments, the finding of a suitable site was a major problem.

After a site has been selected, attention must be given to the moisture content of the near-surface soils. If cohesive soils exist at the site and are partially saturated, steps probably should be taken to saturate the soils if the soils can become saturated at a later date. If the cohesive soils can be submerged in time and if the piles can be subjected to cyclic lateral loads, the site should be flooded during the testing period.

The position of the water table and the moisture content are also important if the soil at the test site is granular. Partial saturation of sand will result in an apparent cohesion that will not be present if the sand dries or if it becomes submerged.

7.3. INVESTIGATION OF SOIL PROPERTIES

In Chapter 2 where procedures are given for obtaining p-y curves, recommendations were made for obtaining soil properties. Those recommendations should be consulted when obtaining data on soils for use in analyzing the results of the lateral-load experiments.

In general, high quality tube samples should be obtained and laboratory tests should be performed on cohesive soils. The Standard Penetration Test is the principal investigative technique for cohesionless soils but the static cone penetrometer test is also recommended.

In performing the soil investigation, careful attention should be given to the near-surface soils, a zone that is frequently given little attention when data are being obtained for design of axially loaded piles. As noted previously, the soil strata within a few diameters of the ground surface provide the principal support for a laterally loaded pile.

7.4. SELECTION OF TEST PILE

If a lateral load test is being performed to confirm the design at a particular site, the diameter, stiffness, and length of the test pile should be as close as possible to similar properties of the piles proposed for production. Because the purpose of the test is to obtain information on soil response, consideration should be given to increasing the stiffness and moment capacity of the test pile in order to allow the test pile to be deflected as much as is reasonable. The additional load that will be necessary will cause no significant problem.

The length of the test pile must be considered with care. As shown in Chapter 1, the deflection of a pile will be significantly greater if it is in the "short" pile range. Tests of these short piles could be hard to interpret because a small change in the pile penetration could cause a large change in the groundline deflection.

The selection of the test pile for the case of complete instrumentation involves a considerable amount of preliminary analysis. Factors to be considered are: the pile diameter for which the soil response is desired, the soil conditions, the kind of instrumentation to be employed for determining bending moment along the length of the pile, the method of installing instrumentation in the pile, the magnitude of the desired groundline deflection, and the nature of the loading. Some examples of piles that were selected for obtaining experimental p-y curves are given in Chapter 2.

7.5 INSTALLATION OF TEST PILE

For cases where information is desired on pile response at a particular site, the installation of the test pile should agree as closely as possible to the procedure proposed for the production piles. It is well known that the response of a pile to load is affected considerably by the installation procedure; thus, the detailed procedure used for pile placement is important.

For the case of a test pile in cohesive soil, the placing of the pile can cause excess porewater pressures to occur. As a rule, these porewater pressures should have dissipated before testing begins; therefore, the use of piezometers at the test site may be important.

The installation of a pile that has been instrumented for the measurement of bending moment along the length of the pile must consider the possible damage of the instrumentation due to pile driving or other installation effects. Pre-boring or some similar procedure may be useful. However, the installation must be such that it is consistent with methods used in practice. In no case would jetting be allowed with wash water flowing up and along the outside of the test pile.

It would be desirable to know how the installation procedures had influenced the soil properties at the test site. However, almost any testing technique would cause soil disturbance and would be undesirable. Some non-intrusive methods are being developed, based on the use of dynamic methods, that can possibly be applied.

Testing of the near-surface soils close to the pile wall at the completion of the load test is useful and can be done without any undesirable effects. The kinds of tests that are desirable are indicated where criteria for p-y curves are discussed. In general, laboratory tests of undisturbed samples are recommended.

7.6. TESTING PROCEDURES

Excellent guidance for the procedures for testing a pile under lateral loading is given by the ASTM Standard D 3966-81, "Standard Method of Testing Piles Under Lateral Loads." Because of its relevance, that document is included in the handbook as Appendix 8. Some general comments on the ASTM standard are given in this section and detailed recommendations are given in the following sections.

For the standard test as well as for the instrumented test, two principles should guide the testing procedure: (1) the loading (static, repeated, sustained, or dynamic) should be consistent with that expected for the production piles and (2) the testing arrangement should be such that deflection, rotation, bending moment, and shear at the groundline (or at the point of load application) are measured or can be computed.

With regard to loading, even though <u>static</u> (short-term) <u>loading</u> is seldom encountered in practice, the soil response from that loading is usually desirable so that correlations can be made with soil properties. As noted later, it is frequently desirable to combine static and <u>repeated loading</u>. A load can be applied, readings taken, and the same load can be re-applied a number of times with readings taken after specific numbers of cycles. Then, a larger load is applied and the procedure repeated. The assumption is made that the readings for the first application at a larger load are unaffected by the repetitions of a smaller load. While that important assumption may not be strictly true, errors are on the conservative side.

As noted earlier, <u>sustained loads</u> will probably have little influence on the behavior of granular materials or on overconsolidated clays if the factor of safety is two or larger (soil stresses are well below ultimate). If a pile is installed in soft, inorganic clay or other compressible soil, sustained loading would obviously influence the soil response. However, loads would probably have to be maintained a long period of time and a special testing program would have to be designed.

The application of a <u>dynamic load</u> to a single pile is feasible and desirable if the production piles sustain such loads. The loading equipment and instrumentation for such a testing program would have to be designed to yield results that would be relevant to a particular application and a special study would be required. The design of piles to withstand the effects of an earthquake involves several levels of computation. Soil-response curves must include an inertia effect and the free-field motion of the earth must be estimated. Therefore, p-y curves that are determined from the tests described herein have only a limited application to the earthquake problem. No method is currently available for performing field tests of piles to gain information on soil response that can be used directly in design of piles to sustain seismic loadings.

The testing of battered piles is mentioned in ASTM D 3966-81. The analysis of a pile group, some of which are batter piles, is discussed in Chapter 6. As shown in that chapter, information is required on the behavior of battered piles under a load that is normal to the axis of the pile. Unless the batter is large, the behavior of battered and vertical piles under this normal load (lateral load) is similar. For large batter, an approxiate solution was given in Chapter 6 that should be adequate in most cases.

The testing of pile groups, also mentioned in D 3966-81, is desirable but is expensive in time, material, and instrumentation. If a large-scale test of a group of piles is proposed, detailed analyses should precede the design of the test in order that measurements can be made that will provide critical information. Such analyses may reveal the desirability of internal instrumentation to measure bending moment in each of the piles.

It is noted in D 3966-81 that the analysis of test results is not covered. The argument can be made, as presented earlier in this chapter, that test results can fail to reveal critical information unless combined with analytical methods. The next section of this chapter suggests procedures that demonstrate the close connection between testing and analysis. A testing program should not be initiated unless preceded and followed by analytical studies.

The ASTM standard mentions methods of dealing with the lateral soil resistance against a pile cap. A conservative procedure, and one that is consistent with reality in many instances, is to assume that there is no soil resistance either against the sides or the bottom of a cap. A small amount of settlement would eliminate the bottom resistance and shrinkage would eliminate the side resistance. Therefore, it is recommended that a pile cap not be used in the testing program or, if used, that the cap not be placed against the soil.

ASTM D 3966-81 gives a number of procedures for applying load and for measuring movements. Some details, generally consistent with the ASTM standard, of methods that have been found to be satisfactory are given in the next section. With regard to a loading schedule, ASTM indicates that loading should be applied in increments to a maximum of 200% to 250% of the design load. Attention is called to a point made earlier: it is rarely possible to perform a test with the rotational restraint at the pile-head exactly the same as for production piles. In view of this fact, an alternate suggestion is made that the loading be continued in increments until the pile actually fails due to the development of a plastic hinge. Or, the loading can be continued until the bending stress becomes equal to a certain percentage of the ultimate, as indicated by computations.

The sections in D 3966-81 on safety requirements and report presentation are worthy of careful consideration. Safety is an important concern in load testing and safety meetings prior to any load test should be held. The detailed list in the section on reporting is useful and indicates most of the items that should be addressed in preparing a report.

7.7. TESTING PILE WITH NO INTERNAL INSTRUMENTATION

A step-by-step procedure is given in the following paragraphs for the testing of a pile or piles with no internal instrumentation, termed the standard test because of its simplicity and ease of performance. The test program is initiated with a study to indicate the economic advantages of the experiment. It is presumed that a careful subsurface investigation with laboratory testing has been carried out and that soil properties are well known. The soil properties near the ground surface are especially important. The particular soil studies that should be made are given in

Chapter 2 where recommendations are given for obtaining p-y curves for various types of soil.

Preliminary Computations

After the type and size of pile has been selected for testing, preliminary computations should be made using Computer Program COM624. The computations should anticipate that the pile head should be free to rotate and that the shear should be applied near the ground surface. Analyses should be done using p-y curves for both static and repeated loading. Curves showing pile-head deflection and pile-head rotation should be developed for a range of loading up to the point where the ultimate moment is developed.

Computations should be done with parameters varied and the length of the test pile and its bending stiffness should be selected on the basis of the computations.

Obtaining Stiffness of Test Pile

The bending stiffness of the test pile or piles can be found by computation but it is preferable to obtain the stiffness experimentally. If the pile consists of a pipe or some other prefabricated section, rather than a cast-in-place pile, it is possible to support the pile near its ends in the laboratory and load it as a beam. The stiffness of the pile can be computed from the deflection.

A method that can be used for a cast-in-place section, or for a prefabricated section as well, is to excavate several feet of soil from around the pile after the primary testing program is completed. The pile is reloaded and deflections are measured at several points along the exposed portion of the pile. If this latter procedure is to be employed, the lateral loading should have been stopped before the pile was damaged.

The stiffness of drilled shafts and other reinforced-concrete sections will vary with bending moment. Some information on this variation can be obtained from the field measurements described above. That information, along with the use of Computer Program PMEIX, should provide engineers with adequate data on stiffness of reinforced-concrete sections.

Pile Installation

As noted earlier, the installation of a test pile should be done in the same manner as for the production piles. Small amounts of accidental batter will have little influence on the performance of a pile under lateral load. Care should be exercised in installation that the near-surface soils have the same properties as for the production piles.

Loading Arrangement

A wide variety of arrangements for the test pile and the reaction system are possible. The arrangement to be selected is the one that has the greatest advantage for the particular design. There are some advantages, however, in testing two piles simultaneously as shown in Fig. 2 of D 3966-81. A reaction system must be supplied and a second pile can supply that need. Furthermore, and more importantly, a comparison of the results of two tests performed simultaneously will give the designer some idea of the natural variations that can be expected in pile performance. It is important to note, however, that spacing between the two piles should be such that the pile-soil-pile interaction is minimized.

Drawings of two two-pile arrangements are shown in Figs. 82 and 83. In both instances the pile head is free to rotate and the loads are applied as near the ground surface as convenient. In both instances free water should be maintained above the ground surface if that situation can exist during the life of the structure.

The details of a system where the piles can be shoved apart or pulled together is shown in Fig. 82. This two-way loading is important if the production piles can be loaded in that manner. The lateral loading on a pile will be predominantly in one direction, termed the forward direction here. If the loading is repeated or cyclic, a smaller load in the reverse direction could conceivably cause the soil response to be different than if the load is applied only in the forward direction. As noted earlier, it is important that the shear and moment be known at the groundline; therefore, the loading arrangement should be designed as shown so that shear only is applied at the point of load application.

Figure 83 shows the details of a second arrangement for testing two piles simultaneously. In this case, however, the load can be applied in only one direction. A single bar of high-strength steel that passes along the diameter of each of the piles is employed in the arrangement shown in Fig. 83a. Two high-strength bars are utilized in the arrangement shown in Fig. 83b. Not shown in the sketches are the means to support the ram and load cell that extend horizontally from the pile. Care must be taken in employing the arrangement shown in either Figs. 82 or 83 to ensure that the loading and measuring systems will be stable under the applied loads.

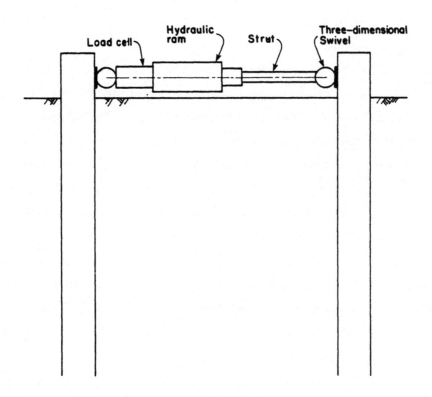

Fig. 82. Two-pile test arrangement for two-way loading

The most convenient way to apply the lateral load is to employ a hydraulic ram with hydraulic pressure developed by an air-operated or electricity-operated hydraulic pump. The capacity of a ram is computed by multiplying the piston area by the maximum pressure. Some rams, of course, are double acting and can apply a forward or reverse load on the test pile or piles. The preliminary computations should ensure that the ram capacity and the piston travel are ample.

If the rate of loading is important (and it may be if the test is in clay soils beneath water and erosion at the pile face is important), the maximum rate of flow of the pump is important along with the volume required per inch of stroke of the ram. The seals on the pump and on the ram, along with hydraulic lines and connections, must be checked ahead of time and spare parts should be available.

High pressures in the operating system constitute a safety problem and can cause operating difficulties. On some projects, the use of an

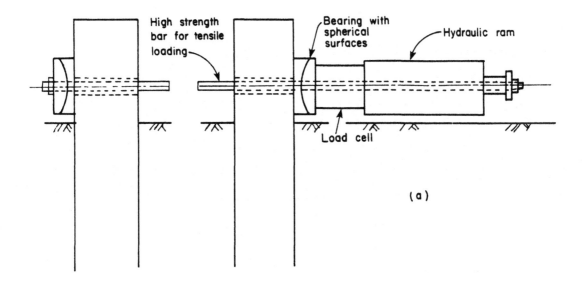

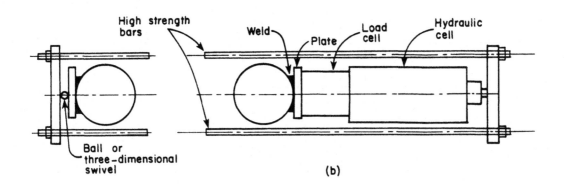

Fig. 83. Two-pile test arrangement with one-way loading

automatic controller for the hydraulic system is justified. A backup control must be available to allow the override of the automatic system in case of malfunction. The writer knows of one important project where the malfunction of the hydraulic system caused a large monetary loss.

The loading system shown in Fig. 83 will ensure that no eccentricity will be applied to the load cell and the hydraulic ram. If the two-bar system shown in Fig. 83 is employed, it should be even simpler to achieve concentric loading. However, the system shown in Fig. 82 will require that the load cell and the ram be attached rigidly together and that bearings be placed at the face of each of the piles so that no eccentric loading is applied to the ram or to the load cell. The arrangement shown in

Fig. 82 may require that the points of load application be adjustable in order to prevent torsional loading of the piles.

Instrumentation

A simple system for obtaining the deflection and rotation of the pile head is shown in Fig. 84. The slope or rotation of the portion of the pile above the point of load application can be found by knowing the gauge readings and the distance between them. The same data will yield the deflection at the point of load application.

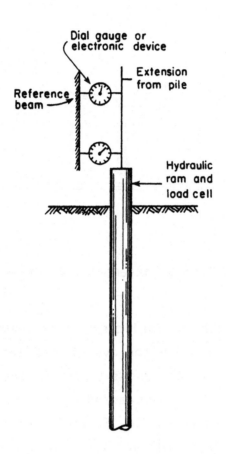

Fig. 84. Schematic drawing of deflection-measuring system

An alternate or redundant method of obtaining the pile-head rotation is shown in Fig. 85. A sensitive bubble for levelling a bar is attached as shown. A sensitive micrometer is fixed to one end of the bar and a hardened point to the other. A sturdy bracket is attached to the pile, or to an extension of the pile, at a convenient distance above the point of load application. Readings of the micrometer when the instrument is carefully levelled for each load that is applied will allow pile-head rotation to be computed.

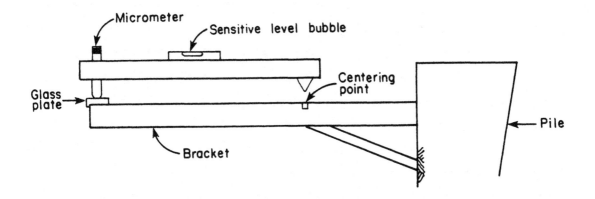

Fig. 85. Device for measuring pile-head rotation

Electronic load cells are available for routine purchase. These cells can be used with a minimum of difficulty and can be tied into a high-speed data-acquisition system if desired.

The motion of the pile head can be measured with dial gauges but a more convenient way is to employ electronic gauges. In either case, gauges with sufficient travel should be obtained or difficulty will be encountered during the test program. Two types of electronic motion transducers are in common use: linear potentiometers or LVDT's (linear variable differential transformers). The LVDT may have a longer life than the differential potentiometer; in either case the motion transducer should be attached so that there is no binding as the motion rod moves in and out.

Two other comments about the instrumentation are important. The verification of the output of each instrument should be an important step in the testing program. Also, the instruments should be checked for temperature sensitivity. In some cases it may be necessary to perform tests at night or to protect the various instruments from all but minor changes in temperature.

Interpretation of Data

The interpretation of data from a test of an uninstrumented pile is a straightforward process. Plots are made of deflection versus applied load and rotation versus applied load (for the groundline or for the point of load application). Computer Program COM624 is then used and computations of pile-head deflection and rotation are made for the same loads that were used for the field test. The results are plotted against the field results. If the results do not agree, the soil parameters (probably the shear strength of clay and angle of internal friction of sand) are changed by trial to bring the computed and experimental results into agreement. (Most of the interpretation will be done in the office; however, it is desirable to do some plotting in the field as a means of checking the validity of the data that are being taken.)

The soil parameters as modified are then used in making a design for the site. An appropriate factor of safety, normally introduced as a load factor to increase the working load, is employed, taking into account the considerations mentioned in Chapter 1.

Example Computation

The test selected for study was performed by Capozzoli (1968) near St. Gabriel, Louisiana. The pile and soil properties are shown in Fig. 86. The loading was short term. The soil at the site was a soft to medium, intact, silty clay. The natural moisture content of the clay varied from 35 to 46 percent in the uppper 10 ft of soil. The undrained shear strength, shown in Fig. 86, was obtained from triaxial tests. The unit weight of the soil was 110 lb/cu ft above the water table and 48 lb/cu ft below the water table.

The results from the field experiment and computed results are shown in Fig. 87. The experimental results are shown by the open circles; the results from Computer Program COM624 with the reported shear strength of 600 lb/sq ft and with an ε_{50} of 1% are shown by the solid line. The soil properties were varied by trial and the best fit to the experimental results was found for an undrained shear strength of 887 lb/sq ft and an ε_{50} of 0.9%. These values of the modified soil properties should be used in design computations for the production piles if the production piles are to be identical with the one employed in the load test.

Computer Program PMEIX was employed and an ultimate bending moment for the section that is shown was computed to be 1392 in-kips. In making

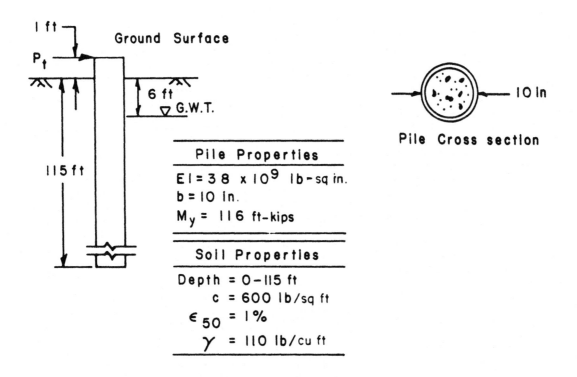

Fig. 86. Information for analysis of test at St. Gabriel

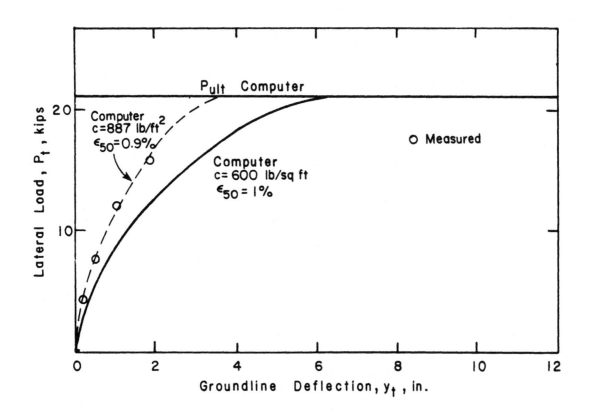

Fig. 87. Comparison of measured and computed results for St. Gabriel Test

the design computations with the modified soil properties, the computed maximum bending moment should be no greater than the ultimate moment (1392 in-kips) divided by an appropriate factor of safety. In computing the maximum bending moment, the rotational restraint at the pile head must be estimated as accurately as possible. If it is assumed that the pile will be unrestrained against rotation and that the load is applied one ft above the groundline, a load of 21 kips will cause the ultimate bending moment to develop. The deflection of the pile must be considered because deflection can control some designs rather than the design being controlled by the bending resistance of the section.

Two other factors must also be considered in design. These are: the nature of the loading and the spacing of the piles. The experiment employed short-term loading; if the loading on the production piles is to be different, an appropriate adjustment must be made in the p-y curves. Also, if the production piles are to be in a closely spaced group, consideration must be given to pile-soil-pile interaction.

7.8. TESTING PILE WITH INTERNAL INSTRUMENTATION

The performance of experiments with piles that are instrumented internally for the measurement of bending moment along the length of the pile is highly desirable. The results of experiments that are carefully performed will allow experimental p-y curves to be developed; thus, significant information can be added to the technical literature. In addition, of course, excellent data will be available to guide the design of piles at the test site. However, the performance of experiments with piles that have internal instrumentation is expensive, both in labor and materials. In addition, instrumentation specialists with excellent skills are required. Therefore, a detailed cost-benefit study should be undertaken before such a test program is begun.

Preliminary Computations

If a major experiment with a pile with internal instrumentation is to be undertaken, the preliminary computations should be exhaustive. Assuming that the test site and the pile geometry have been selected and that soil properties are known, computations must be performed to get the best possible estimate of the response of the pile. On the basis of these compuations, the nature of the loading system is decided upon and a detailed design of the system is carried out.

The preliminary computations also allow the selection of the kind of internal instrumentation that is to be employed and a detailed design of the instrumentation is then made. Electrical-resistance strain gauges are frequently employed to read strains in the pile material. The test pile can be calibrated by supporting the pile as a beam, applying known measurements at positions of strain gauges, and reading the output of each gauging point. If a drilled shaft is employed in the test, an instrumented pipe can be cast along the axis of the shaft and calibration can be done after the test is over by removing soil around the pile to as great a depth as possible and reloading the pile (Welch and Reese, 1972).

Further computations must be carried out to ensure that the pile is not damaged if it is to be installed by driving. Diligence in planning and in performing preliminary computations can do much to assure the success of the expensive instrument.

Instrumentation

The instrumentation that is placed above the ground is similar, if not identical, to that described for the pile with no internal instrumentation. While the principal item of internal instrumentation pertains to a direct determination of bending moment from point to point along the pile, the use of a slope indicator from which deflections can be obtained is sometimes desirable. If space allows and if the loading schedule that is proposed will allow a slope indicator to be used, the installation of slope-indicator casing may be warranted.

As noted above, the use of strain gauges to enable bending moments to be obtained is a common practice. However, innovative techniques are being developed regularly and the selection and installation of the internal instrumentation should follow a careful study of available methods.

Some investigators have made measurements of ground-surface movements during the lateral loading of a pile. The placing of markers on a grid pattern around the test pile and the measurement of the movement of those markers are time-consuming and cumbersome. The use of photographic techniques to obtain ground-surface movements has much to recommend it.

Analysis of Data and Correlations with Theory

The principal analytical technique is to perform two integrations of the bending moment curves and two differentiations. The boundary conditions at the head of the pile must be employed in the analysis. The integrations yield the pile deflections; with reasonably good moment curves and with good measurements of the boundary conditions at the pile head, an accurate family of curves giving deflection of the pile as a function of depth can be obtained.

The two differentiations are another matter. Errors in the moment values are accentuated. Therefore, it is usually necessary to employ curve-fitting techniques and obtain analytical expressions for selected portions of the moment curves. If the differentiations can be carried out successfully, the result will be a family of curves showing soil resistance as a function of depth. Specific depths can be selected and cross-plotting will yield a family of p-y curves.

An additional step in the analytical process is to employ the principles of soil mechanics and of elasticity to develop predictive equations for pile response. Ideally, the predictive equations should agree with the experimental results at the test site and should further serve to predict the behavior of piles of different geometry at the test site and at other sites where the soils are similar. The predictive equations will be valid, of course, only for the kind of loading employed at the test site.

Review of Experiments Using Piles with Internal Instrumentation

Matlock (1970) performed experiments near Austin, Texas, and near Sabine, Texas, in soft to medium clay. The pile was a steel pipe, 12.75 inches in diameter. Thirty-five pairs of electrical-resistance strain gauges were installed in the interior of the pipe. The gauges were spaced 6 in apart near the top of the embedded portion with wider spacings being used below. The embedded portion of the pile was 45 ft long. The pipe was split along a diameter, the gauges were installed, and the two halves were welded together.

The pile was calibrated prior to driving so that extremely accurate determinations of bending moment could be made. The experimental p-y curves that were obtained from the testing program form the basis of recommendations that are widely used for design of piles in soft clay under lateral loading.

Cox, Reese, and Grubbs (1974) performed experiments near Corpus Christi, Texas, in sand. The piles were steel pipes, 24 inches in diameter. Forty electrical-resistance strain gauges were installed in each of two piles by placing the piles horizontally and by working from a trolley. Two piles were driven at the same site; one pile was tested under static loading and the other under cyclic loading. The embedded length of each pile was 69 ft.

The piles were calibrated in the laboratory prior to installation. The experimental p-y curves that were obtained from the testing program form the basis of recommendations that are widely used for the design of piles in sand under lateral loading.

7.9. CONCLUDING COMMENTS

Only a brief presentation is possible concerning the details of a program of testing of piles under lateral load. The brevity of the presentation is consistent with the purposes of the handbook and is not meant to detract from the importance of the topic.

Simple, inexpensive experiments can be performed with piles with no internal instrumentation and data of great value can be obtained concerning the response of a pile at a particular site.

The performance of tests of piles with internal instrumentation can well be justified at the site of a major project, especially if the current methods of predicting p-y curves are not exactly applicable to the soil, pile, and loading to be employed at the site. In addition to getting data for the design of a particular project, data will be obtained for use at similar sites. Also, a contribution can be made to the engineering profession.

Redundancy in load-measuring and deflection-measuring systems is good practice. Rams can be calibrated as a means of checking readings from load cells. Stretched wires or surveying instruments can be employed to check deflection. Such redundancy can be extremely useful in case of the failure of a primary system of measurement.

The available data are insufficient to allow a comment to be made that all field tests of piles under lateral loading are cost effective. However, the tests that have been performed appear to have saved money on specific projects. The tests of instrumented piles have paid for themselves many times over. The investigation of the benefits from performing field tests of piles under lateral loading for a specific project is strongly advised.

APPENDIX 1

INFORMATION ON INPUT OF DATA FOR COMPUTER PROGRAM COM624

Note: The information in this appendix has been abstracted from "Documentation of Computer Program COM624," by Lymon C. Reese and W. Randall Sullivan, Geotechnical Engineering Center, Bureau of Engineering Research The University of Texas at Austin, August, 1980

APPENDIX 1

INFORMATION ON INPUT OF DATA FOR COMPUTER PROGRAM COM624

A1.1 INTRODUCTORY REMARKS

Data input is based on a coordinate system in which the pile head (top of pile) is the origin and the positive x-direction is downward (Fig. A1.1). The ground surface need not be at the elevation of the pile head. Sign conventions are shown in Fig. A1.2.

The program is organized so that up to 50 problems can be analyzed in a single run; this facilitates sensitivity studies of input variables with minimum effort by the user.

Any convenient and consistent units of force and length can be used. The program is set up to label the output in one of three ways:

1. The user can designate that English units of inches and pounds will be used, and output will be labeled accordingly;
2. The user can specify that metric units of kilonewtons and meters will be used, and output will be labeled accordingly;
3. The user can use any other consistent units (the computer does not need to know which ones) of force and length, and output will be labeled in terms of forces and lengths (F and L).

Several default values may be used in data input. Where the user desires to use a default value, he should leave the input blank for the relevant variable.

In this section, the symbol "O" denotes a capital letter O; the symbol "Ø" denotes the number zero.

A1.2 PREPARATION FOR INPUT

The following steps are recommended to prepare for data input.
1. Decide which units will be used for force and length.
2. Decide into how many increments the pile is to be divided. Up to 3ØØ increments are allowed. Be sure to satisfy your-

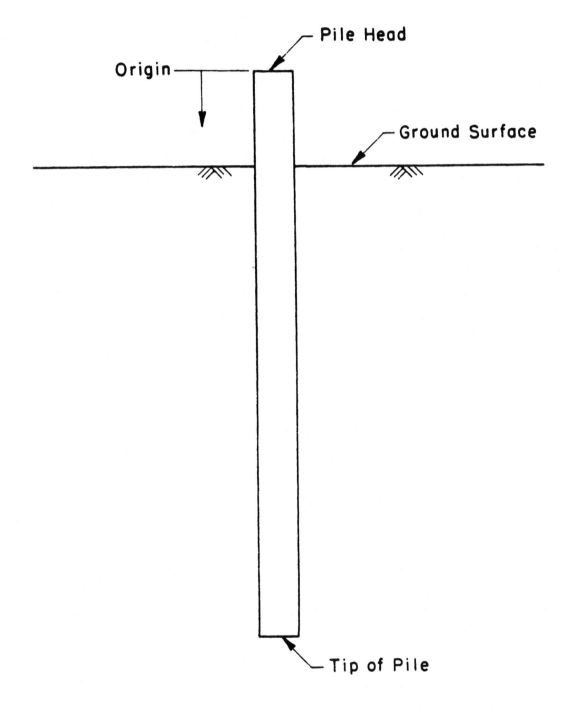

Fig. A1.1. Coordinate system

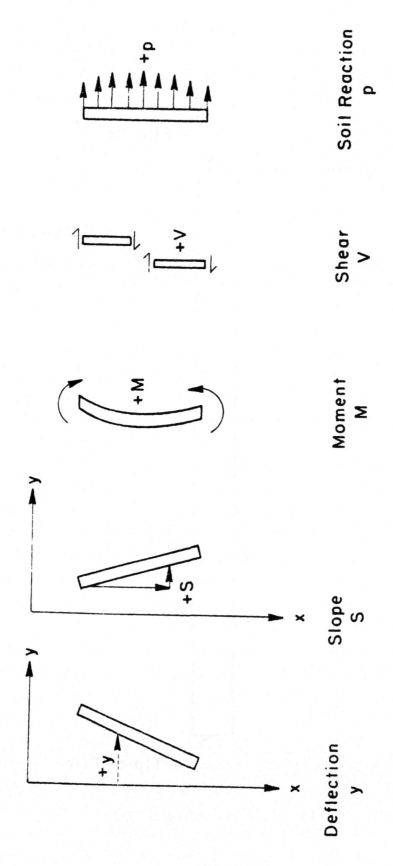

A1.2. Sign convention

self that an adequate number of increments have been used to obtain a satisfactory solution.

3. Decide whether p-y curves will be input or whether they will be generated internally. If they are to be input, pick depths for input, pick the number of points to be input for each depth, and tabulate the data. Up to 30 p-y curves are allowed.

4. If p-y curves are to be generated internally, divide the soil profile into from one to nine layers; decide which of the following p-y criteria will be used for each layer:
- Matlock's (1970) criteria for soft clay;
- Reese et al.'s (1975) criteria for stiff clay below the water table:
- Welch and Reese's (1975) criteria for stiff clay above the water surface;
- Reese et al.'s (1974) criteria for sand;
- Sullivan et al.'s (1979) uniform criteria for clay.

Estimate undrained shear strength c and strain at 50 percent stress level ε_{50} for clay layers; estimate the angle of internal friction ϕ for sand; and estimate the slope k of a plot of maximum soil modulus E_s versus depth x for all strata. If unified clay criteria are used, determine values of A and F to be used. Be sure to read the background on any criteria that are used so that you thoroughly understand their basis and limitations.

5. Note the length of the pile, the modulus of elasticity of pile material, and the x-coordinate of the ground surface.

6. Divide the pile into from one to ten segments with uniform cross-section. For each segment, tabulate the x-coordinate of the top of the segment, the diameter of the segment, the moment of inertia, and, unless the pile is a pipe section, the area of the pile.

7. If there are any distributed lateral loads on the pile, tabulate up to ten points on a plot of distributed load versus depth below top of pile.

8. Tabulate up to ten points on a plot of effective unit weight of soil versus depth. This step is not necessary if no p-y curves are generated internally.

9. Tabulate up to ten points on plots of c, ϕ, and ε_{50} versus x. Skip this step if no p-y curves will be generated internally.

10. If p-y curves are generated internally in the program, tabulate any depths for which p-y curves are to be printed. Ordinarily, a few curves are printed for verification purposes.

11. Determine the loads to the top of the pile;
    A. Lateral load at pile head
    B. Second boundary condition at pile head, which can be either
        i. moment (M_t)
        ii. slope (S_t)
        iii. rotational restraint (M_t/S_t).
    C. Axial load (assumed to be uniform over full length of pile) Up to 20 loading combinations can be input for each problem, e.g., to generate a load deflection curve.

A1.3 LINE-BY-LINE INPUT GUIDE (See Appendix A1.6 for the input form.)

Title Card

    Variable: TITLE(I)

    Format: 18A4

    Number of Cards: 1

    Explanation: Any characters, including blanks, are allowed in this descriptive title. However, do not type the word END in columns 1 through 3 as this is used to indicate the end of the data input.

Units Card

    Variables: ISYSTM, IDUM1, IDUM2, IDUM3

    Format: 4A4

    Number of Cards: 1

    Explanation: In columns 1 through 4, type:

        ISYSTM = ENGL if English units of pounds and inches are to be used;

= METR if metric units of kilonewtons and meters are to be used;

= Anything else if some other consistent set of units for force and length are to be used (the program will not try to determine which set of units is used but will indicate units on output by F for force and L for length, e.g., stress would be F/L**2).

IDUM1-3 = Any characters (in columns 5-16) to describe the system of units selected.

Input Control Card

    Variables: NI, NL, NDIAM, NW

    Format: 4I5

    Number of Cards: 1

    Explanation: NI = number of increments into which the pile is divided (maximum is 300)

    NL = number of layers of soil (maximum is 9)

    NDIAM = number of segments of pile with different diameter, area, or moment of inertia (maximum is 10)

    NW = number of points on plot of distributed lateral load on the pile versus depth (minimum is 0, maximum is 10).

NDIAM must be > 0 for the first problem in the data deck. If NDIAM = 0 for subsequent problems, the same pile properties used in the previous problem will be used again in the subsequent problem.

Set NW = 0 if there are no distributed loads on the pile. Set NW = -1 for the second or any subsequent problem in a data deck if you want the same distributed loads to be used again.

Set NL = 0 if the same soil profile is to be used as was used in the previous problem in the data deck.

Input Control Card

    Variables: NG1, NSTR, NPY

    Format: 3I5

    Number of Cards: 1

Explanation: NG1 = number of points on plot of effective unit weight versus depth (minimum = 2, maximum = 10)

NSTR = number of points on input curves of strength parameters (c, ϕ, ε_{50}) versus depth (minimum = 2, maximum = 10)

NPY = number of input p-y curves (minimum = 0, maximum = 30)

NG1 may equal 0 for the second or any subsequent problem in a data deck if the same unit weight plot is to be used as was used in the previous problem.

Set NPY = -1 in the second or any subsequent problem in a data deck to retain the input p-y curves from the previous problem and therefore to avoid re-reading the data.

Set NG1 = 0 and NSTR = 0 if all p-y curves are to be input by the user (if no p-y curves are to be generated internally).

Geometry Card

Variables: LENGTH, EPILE, XGS

Format: 3E10.3

Number of Cards: 1

Explanation: LENGTH = length of pile (L)

EPILE = Modulus of elasticity of pile (F/L²)

XGS = depth below top of pile to ground surface (L)

Output Control Card

Variables: KPYOP, INC

Format: 2I5

Number of Cards: 1

Explanation: KPYOP = 0 if no p-y curves are to be generated and printed for verification purposes

= 1 if p-y curves are to be generated and printed for verification (see "control card for output of Internally-Generated p-y curves" card for input of depths at which p-y curves will be generated and printed)

INC = increment used in printing output

= 1 to print values at every node

= 2 to print values at every second node

= 3 to print values at every third node, etc. (up to NI + 1).

Any p-y curves generated for output are written to TAPE1.

Analysis Control Card

Variables: KBC, KOUTPT, KCYCL, RCYCL

Format: 3I5, E10.3

Number of Cards: 1

Explanation: KBC = code to control boundary condition at top of pile

= 1 for a free head (user specifies shear P_t and moment M_t at the pile head)

= 2 for specified shear P_t and slope S_t at the pile head (S_t = 0 for a fixed-head pile)

= 3 for a specified shear P_t and rotational restraint M_t/S_t at the pile head.

KOUTPT = 0 if data are to be printed only to depth where moment first changes sign

= 1 if data are to be printed for full length of pile

= 2 for extra output to help with debugging

KCYCL = 0 for cyclic loading

= 1 for static loading

RCYCL = number of cycles of loading (need only for p-y curves generated with criteria for stiff clay above the water table). Default = 100.0.

Run Control Card

Variables: MAXIT, YTOL, EXDEFL

Format: I5, 2E10.3

Number of Cards: 1

Explanation: MAXIT = Maximum number of iterations allowed for analysis of single set of loads. Leave blank for default value of 100 to be used.

YTOL = tolerance (L) on solution convergence. When the maximum change in deflection at any node for successive iterations is less than YTOL, iteration stops. Leave blank for default value of 1.0 E -5 to be used.

EXDEFL = value of deflection of pile head (L) that is considered grossly excessive and which stops the run. Leave blank for a default value equal to ten times the diameter of the top of the pile.

Distributed Loads

Omit if NW = 0 or NW = -1.

    Variables: XW(I), WW(I)

    Format: 2E10.3

    Number of Cards: NW

    Explanation: XW = depth (L) below top of pile to a point where distributed load is specified

                   WW = distributed lateral load (F/L) on pile

The program uses linear interpolation between points on the WW-XW curve to determine the distributed load at every node. For best results, points on the WW-XW curve should fall on the pile node points. Wherever no distributed load is specified, it is assumed to be zero. Data must be arranged with ascending values of XW.

Pile Properties Card

Omit if NDIAM = 0.

    Variables: XDIAM(I), DIAM(I), MINERT(I), AREA(I)

    Format: 4E10.3

    Number of Cards: NDIAM

    Explanation: XDIAM = x-coordinate (depth below top of pile) of the top of a segment of pile with uniform cross-section (L). The first depth (XDIAM (1)) must equal 0.0.

                   DIAM = diameter of pile corresponding to XDIAM (L). For non-circular cross-sections, use of minimum width will produce conservative results.

MINERT = moment of inertia of pile cross-section (L^4)

AREA = cross-sectional area of pile (L^2). If left blank, program will compute area assuming a pipe section.

Data must be arranged with ascending values of XDIAM. Note that at a depth between XDIAM(I) and XDIAM(I + 1), the pile properties associated with XDIAM(I) will be used. For a pile with uniform cross-section, just one pile property card is needed. The last value of XDIAM need not be greater than or equal to the length of pile.

<u>Soil Profile Card</u>

Omit this card if NL = 0.

Variables: LAYER, KSOIL, XTOP, XBOT, K, AE, FR

Format: 2I5, 5E10.3

Number of Cards: NL

Explanation: LAYER(I) = layer identification number (use 1 for the top layer, 2 for the second layer, etc.)

KSOIL(I) = code to control the type of p-y curves that will be used for L-th layer

= 1 to have p-y curves computed internally using Matlock's (1970) criteria for soft clay

= 2 to have p-y curves computed internally using Reese et al.'s (1975) criteria for stiff clay below the water table

= 3 to have p-y curves computed internally using Reese and Welch's (1975) criteria for stiff clay above the water table

= 4 to have p-y curves computed internally using Reese et al.'s (1974) criteria for sand

= 5 to use linear interpolation between input p-y curves

      = 6 to have p-y curves computed internally using Sullivan et al.'s (1979) unified clay criteria

  XTOP(I) = x-coordinate of top of layer (L)

  XBOT(I) = x-coordinate of bottom of layer (L)

  K(I) = constant (F/L^3) in equation E_s = kx. This is used (1) to define initial soil moduli for the first iteration and (2) to determine initial slope of p-y curve where KSOIL = 2, 4, or 6.

  AE(I) = Factor "A" in uniform clay criteria (leave blank unless KSOIL = 6)

  FR(I) = Factor "F" in uniform clay criteria (leave blank unless KSOIL = 6)

Arrange data in ascending order of LAYER(I).

Unit Weight Card

Omit this card if NG1 = 0.

  Variables: XG1(I), GAM1(I)

  Format: 2E10.3

  Number of Cards: NG1

  Explanation: XG1 = depth below top of pile to point where effective unit weight of soil is specified (L)

  GAM1 = effective unit weight of soil(F/L^3) corresponding to XG1

The first depth (XG1(I)) must not be greater than the x-coordinate of the ground surface and the last depth (XG1(NG1)) must not be less than the length of pile. The program interpolates linearly between points on XG1 - GAM1 curve to determine effective unit weight of soil at a particular depth. The data must be arranged with ascending values of XG1.

Strength Parameter Card

Omit this card if NSTR = 0.

  Variables: XSTR(I), C1(I), PHI1(I), EE50(I)

  Format: 4E10.3

  Number of Cards: NSTR

  Explanation: XSTR = x-coordinate (depth below top of pile) for which c, ϕ, and ε_{50} are specified (L)

C1 = undrained shear strength of soil (F/L²) corresponding to XSTR

PHI1 = angle of internal friction (ϕ, in degrees) corresponding to XSTR

EE50 = strain at 50 percent stress level (ε_{50}, dimensionless) corresponding to XSTR

The program uses linear interpolation to find c, ϕ, and ε_{50} at points between input XSTR's. XSTR(I) should not be greater than the x-coordinate of the ground surface and XSTR(NSTR) should not be less than the length of the pile. Arrange data with ascending values of XSTR. For clay layers (KSOIL = 1, 2, 3, or 6), PHI1 will not be used and may be left blank. For sand layers (KSOIL = 4), C1 and EE50 are not used and may be left blank.

Control Card for Input of p-y Curves

Omit this card if NPY = 0 or NPY = -1.

    Variable: NPPY

    Format: I5

    Number of Cards: 1

    Explanation: NPPY = number of points on input p-y curves (minimum = 2, maximum = 30)

Card for Depth of p-y Curve

Omit this card if NPY = 0 or NPY = -1.

    Variable: XPY(I)

    Format: E10.3

    Number of Cards: 1

    Explanation: XPY = x-coordinate (depth below top of pile) to an input p-y curve (L)

Data must be arranged in ascending order of XPY. Input XPY, then data to define the associated p-y curve (see next card), then the next XPY, etc.

p-y Curve Data Card

Omit if NPY = 0 or NPY = -1.

    Variables: YP(I,J), PP(I,J)

    Format: 2E10.3

    Number of Cards: NPY * NPPY

    Explanation: YP = deflection (L) of a point on a p-y curve

                       PP = soil resistance (F/L) corresponding to YP

Data must be arranged in ascending order of YP. Sequence of input is as follows:

```
      DO 30 I=1, NPY
      READ (5,10), XPY(I)
   10 FORMAT (E10.3)
      READ (5,20), (YP(I,J), PP(I,J), J=1, NPPY)
   20 FORMAT (2E10.3)
   30 CONTINUE
```

The program interpolates linearly between points on a p-y curve and between p-y curves. The program uses the deepest p-y curve available for any nodes that extend below the depth of the deepest p-y curve.

<u>Control Card for Output of Internally-Generated p-y Curves</u>

Omit this card if KPYOP = 0

    Variable: NN

    Format: I5

    Number of Cards: 1

    Explanation: NN = number of depths for which internally-generated p-y curves are to be printed (maximum = 305).

Internally-generated p-y curves may be computed for selected depths and printed for verification purposes. In the analysis of pile response, a separate p-y curve is calculated at every node. Therefore, the number of p-y curves printed will have no effect on the solution.

<u>Control Card for Depths at Which Internally-Generated p-y Curves are to be Printed</u>

Omit this card if KPYOP = 0.

    Variable: XN(I)

    Format: E10.3

    Number of Cards: NN

    Explanation: XN = X - coordinate (L) at which internally-generated p-y curves are to be generated and printed.

<u>Card to Establish Loads on Pile Head</u>

    Variables: KOP, PT, BC2, PX

    Format: I5, 3E10.3

    Number of Cards: Between 1 and 20

Explanation: KOP = 0 if only the pile head deflection, slope, maximum bending moment, and maximum combined stress are to be printed for the associated loads.

= 1 if complete output is desired for the associated loads

= -1 to indicate that all pile head loads have been read and to terminate reading this card.

PT = lateral load (F) at top of pile

BC2 = value of second boundary condition

= moment (F-L) at top of pile if KBC = 1

= slope (dimensionless) at top of pile if KBC = 2

= rotational stiffness (F-L), or moment divided by slope, if KBC = 3

PX = axial load (F) on pile (assumed to be uniform over whole length of pile)

Set KOP = -1 to stop input of loads on pile head.

<u>Card to Stop Run</u>

Variable: TITLE(I)

Format: 18A4

Explanation: TITLE = END to stop reading data.

This is the descriptive title for the run (see explanation of first card in the data deck). If the word END is typed in columns 1-3, and column 4 is blank, the program will stop. If anything else appears in these columns, the card will be assumed to be a descriptive title for a new problem. Up to 50 problems can be analyzed in one run. If a new problem is to be read, return to the beginning of this input guide to read the title card and further data. Input is identical no matter what problem is analyzed, except that on second and subsequent problems, some parameters (NL, NDIAM, NW, NG1, NPY) can be set equal to zero (or in some cases -1), to avoid inputing redundant data.

A1.4 SUMMARY

A flowchart for input of data is presented in Fig. A1.3. A summary of input variables and a dictionary of variables are shown in Tables A1.1 and A1.2, respectively.

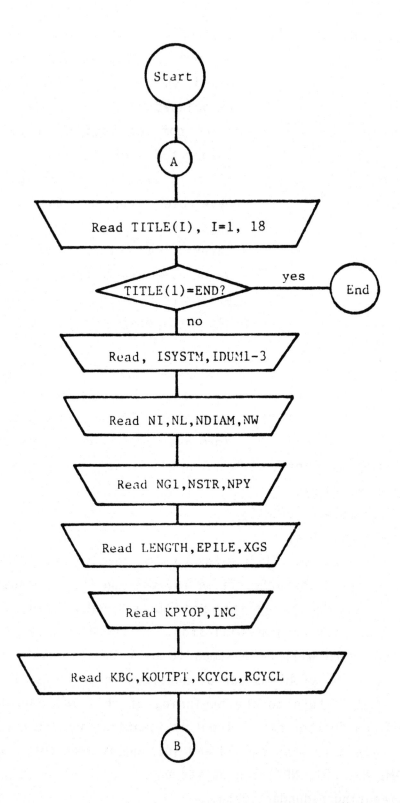

Fig. A1.3. Flowchart for input to program COM624

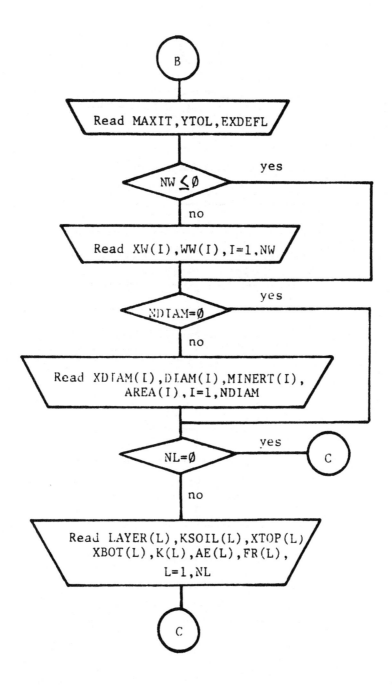

Fig. A1.3. Flowchart for input to program COM624 (cont.)

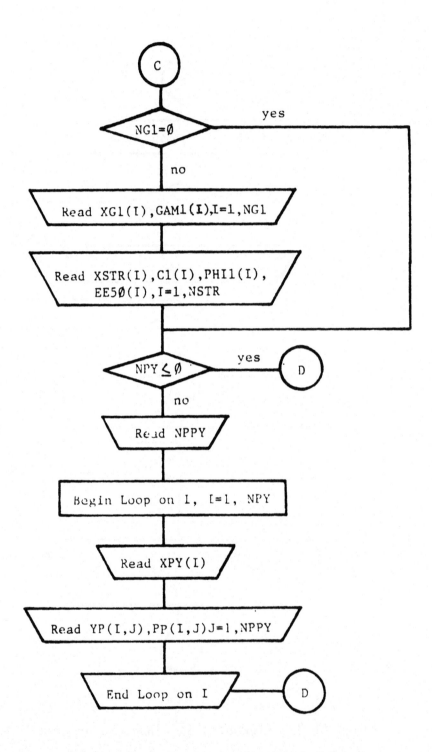

Fig. A1.3. Flowchart for input to program COM624 (cont.)

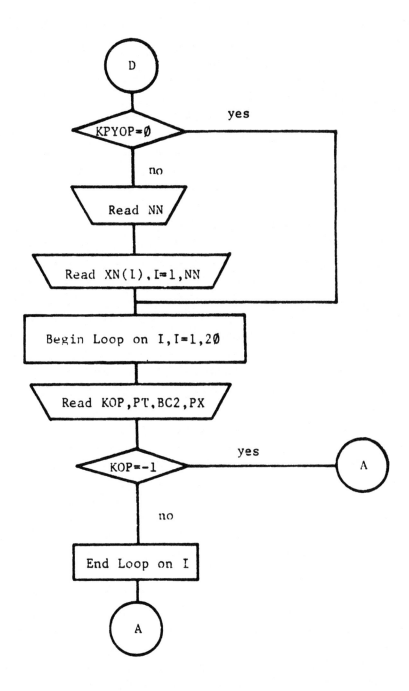

Fig. A1.3. Flowchart for input to program COM624 (cont.)

TABLE A1.1. SUMMARY OF INPUT VARIABLES AND FORMATS

Card	Number of Cards	Variables (including dimensions)	Format
A	1	TITLE(18)	18A4
B	1	ISYSTM,IDUM1,IDUM2,IDUM3	4A4
C	1	NI,NL,NDIAM,NW	4I4
D	1	NG1,NSTR,NPY	3I5
E	1	LENGTH,EPILE,XGS	3E10.3
F	1	KPYOP,INC	2I5
G	1	KBC,KOUTPT,KCYCL,RCYCL	3I5,E10.3
H	1	MAXIT,YTOL,EXDEFL	I5,2E10.3
I	NW	XW(10),WW(10)	2E10.3
J	NDIAM	XDIAM(10),DIAM(10),MINERT(10),AREA(10)	4E10.3
K	NL	LAYER(10),KSOIL(10),XTOP(10),XBOT(10),K(10),AE(10),FR(10)	2I5,5E10.3
L	NG1	XG1(10),GAM1(10)	2E10.3
M	NSTR	XSTR(10),C1(10),PHI1(10),EE50(10)	4E10.3
N	0 or 1	NPPY	I5
O	NPY	XPY(30)	E10.3
P	NPPY	YP(30,30),PP(30,30)	2E10.3
Q	0 or 1	NN	I5
R	NN	XN(305)	E10.3
S	1 to 20	KOP,PT,BC2,PX	I5,3E10.3

TABLE A1.2. DICTIONARY - INPUT VARIABLES

Program Variable	Type	Size	Analysis Variable	Description
TITLE	Integ.	18	–	Descriptive title of problem; TITLE(1)=END (beginning in column 1) to indicate that all data have been input.
ISYSTM	Integ.	1	–	ISYSTM=ENGL for use of English units of pounds and inches; =METR for metric units of kilonewtons and meters; =anything else for a user-selected consistent set of units for force and length.
IDUM1-3	Integ.	1	–	Description of units selected (to appear with output).
NI	Integ.	1	–	Number of increments of pile (maximum=300).
NL	Integ.	1	–	Number of soil layers (maximum=9).
NDIAM	Integ.	1	–	Number of pile segments with different cross-sections.
NW	Integ.	1	–	Number of points on plot of distributed lateral load vs. depth (maximum=10). NW=0 if there are no distributed loads; NW=-1 to use same distributed loads as in the last problem in the run.
NG1	Integ.	1	–	Number of points on plot of effective unit weight of soil vs. depth (maximum=10).
NSTR	Integ.	1	–	Number of points on plots of strength properties of soil vs. depth (maximum=10).

TABLE A1.2. DICTIONARY - INPUT VARIABLES (cont.)

Program Variable	Type	Size	Analysis Variable	Description
NPY	Integ.	1	-	Number of input p-y curves (maximum=30). NPY=0 if no p-y curves are to be input; NPY=-1 if same p-y curves as used in last problem are to be used again in current problem.
LENGTH	Real	1	-	Length of pile (L).
EPILE	Real	1	E	Modulus of elasticity of pile material (F/L^2).
XGS	Real	1	-	Depth below top of pile to the ground surface (L).
KPYOP	Integ.	1	-	KPYOP=0 if no internally-generated p-y curves are to be printed; =1 if internally-generated p-y curves are to be printed for depths specified by the user.
INC	Integ.	1	-	Increment for printing results; INC=1 to print results for every node; =2 to print results for every second node; =3 to print results for every third node, etc.
KBC	Integ.	1	-	Code to control boundary condition at pile head; KBC=1 for a free head; =2 for a specified rotation; =3 for a specified rotational restraint.
KOUTPT	Integ.	1	-	Code to control level of output; KOUTPT=0 to print table of depth-deflection-moment only to the depth where moment first changes sign (for free head case only);

TABLE A1.2. DICTIONARY - INPUT VARIABLES (cont.)

Program Variable	Type	Size	Analysis Variable	Description
KOUTPT (cont.)	Integ.	1	-	=1 to print table for full length of pile; =2 to obtain additional output on final iteration for debugging purposes.
KCYCL	Integ.	1	-	KCYCL=0 for cyclic loading; =1 for static loading.
RCYCL	Real	1	-	Number of cycles of loading (needed only if KSOIL=3 for at least 1 layer).
MAXIT	Integ.	1	-	Maximum number of iterations allowed.
YTOL	Real	1	-	Tolerance on solution convergence (L)=maximum allowable differences in deflection at any node for two successive iterations.
EXDEFL	Real	1	-	Value of deflection that is considered grossly excessive and which will stop the analysis (L).
XW	Real	10	-	x-coordinate at which distributed load is to be specified (L).
WW	Real	10	-	Distributed load corresponding to XW (F/L).
XDIAM	Real	10	-	x-coordinate of the top of a segment of pile with uniform cross-section (L).
DIAM	Real	10	d	Diameter of pile corresponding to XDIAM (L).
MINERT	Real	10	I	Moment of inertia of pile cross-section corresponding to XDIAM (L^4).

TABLE A1.2. DICTIONARY - INPUT VARIABLES (cont.)

Program Variable	Type	Size	Analysis Variable	Description
AREA	Real	10	A	Cross-sectional area of pile corresponding to XDIAM (L^2).
LAYER	Integ.	10	-	Identification number for soil layer.
KSOIL	Integ.	10	-	Code to control type of p-y curves that are to be used; KSOIL=1 to have curves generated internally using criteria for soft clay; =2 to have curves generated internally using criteria for stiff clay below the water table; =3 to have curves generated internally using criteria for stiff clay above the water table; =4 to have curves generated internally using criteria for sand; =5 to use p-y curves input by the user; =6 to use p-y curves generated internally using unified clay criteria.
XTOP	Real	10	-	x-coordinate of top of a soil layer (L).
XBOT	Real	10	-	x-coordinate of bottom of a soil layer (L).
K	Real	10	k	Slope of plot of soil modulus (E_s) vs. depth (x), F/L^3.
AE	Real	10	A	Factor in unified clay criteria.
FR	Real	10	F	Factor in unified clay criteria.
XG1	Real	10	-	x-coordiante at which effective unit weight of soil is specified (L).

TABLE A1.2. DICTIONARY - INPUT VARIABLES (cont.)

Program Variable	Type	Size	Analysis Variable	Description
GAM1	Real	10	γ	Effective unit weight of soil corresponding to GAM1 (F/L^3).
XSTR	Real	10	-	x-coordinate at which strength properties of soil are specified (L).
C1	Real	10	c	Undrained shear strength of soil (F/L^2).
PHI1	Real	10	ϕ	Angle of internal friction of soil (degrees).
EE50	Real	10	ε_{50}	Strain at 50 percent stress level of soil (dimensionless).
NPPY	Integ.	1	-	Number of points used to define an input p-y curve.
XPY	Real	30	-	x-coordinate of an input p-y curve (L)
YP	Real	30,30	y	Deflection of input p-y curve (L).
PP	Real	30,30	p	Soil reaction input on p-y curve (F/L).
NN	Integ.	1	-	Number of depths at which p-y curves are to be generated for output.
XN	Real	305	-	Depth at which p-y curve is to be generated for output (L)
KOP	Integ.	1	-	KOP=0 if only the pile head deflection, slope, maximum bending moment, and maximum combined stress in the pile are to be printed for a particular load; =1 for complete output for

TABLE A1.2. DICTIONARY - INPUT VARIABLES (cont.)

Program Variable	Type	Size	Analysis Variable	Description
KOP (cont.)	Integ.	1	-	the load; =-1 to indicate that all loading conditions to be analyzed have been input.
PT	Real	1	P_t	Lateral load at top of pile (F).
BC2	Real	1	-	Value of second boundary condition; BC2=moment at pile head if KBC=1 (F-L); =slope at pile head if KBC=2 (dimensionless); =rotational stiffness at pile head if KBC=3 (F-L).
PX	Real	1	P_x	Axial load on pile (F).

A1.5 ERROR MESSAGES

Excessive Deflection

If deflection at the pile head in any iteration exceeds EXDEFL, the following message is printed:

> * * * * * FATAL ERROR * * * * *
> DEFLECTION AT PILE HEAD = _____
> IS LARGER THAN ALLOWABLE VALUE OF _____
> AFTER _____ ITERATIONS

where the blanks are filled in with appropriate numbers. Cause of this problem might be excessive loads, errors in p-y curves, values of k (in equation $E_s = kx$) that are too small, or too small a value of EXDEFL. The program will skip to next load if this problem is encountered.

Convergence

If the solution fails to converge within permitted number of iterations, the following message is printed:

> * * * * * WARNING * * * * *
> SOLUTION DID NOT CONVERGE
> MAXIMUM DEFLECTION ERROR = _____

where the blank is filled in with appropriate number. The cause of this problem might be too few iterations allowed (MAXIT too small) or too small a tolerance for solution convergence (YTOL too small). Program sill skip to next load if this problem is encountered.

Too Many Loads

If the user tries to read more than 20 loadings in one analysis, the following message will be printed:

> * * * * * FATAL ERROR * * * * *
> YOU TRIED TO INPUT MORE THAN 20
> LOADS. PROGRAM STOPPED.

This error causes execution of the program to end. Cause of problem is too many loads or forgetting to set KOP = -1 to stop reading loads.

Soil Profile Not Deep Enough

If the length of the pile exceeds the x-coordinate of the bottom of the deepest layer, the following message is printed:

> * * * * * FATAL ERROR * * * * *
> THE PILE LENGTH, _____, EXTENDS
> BELOW THE BOTTOM OF THE DEEPEST
> LAYER.

This error will cause the program to stop. To correct the problem, be sure the bottom of the deepest layer extends below the tip of the pile.

A1.6 CODING FORM FOR COM624

(See following pages.)

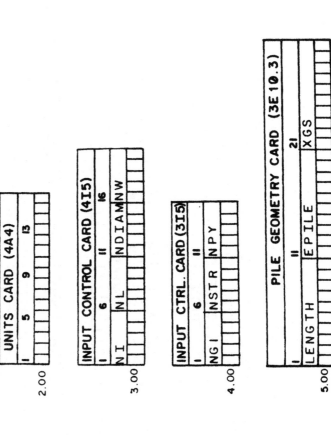

Coding Form for COM624

Coding Form for COM624

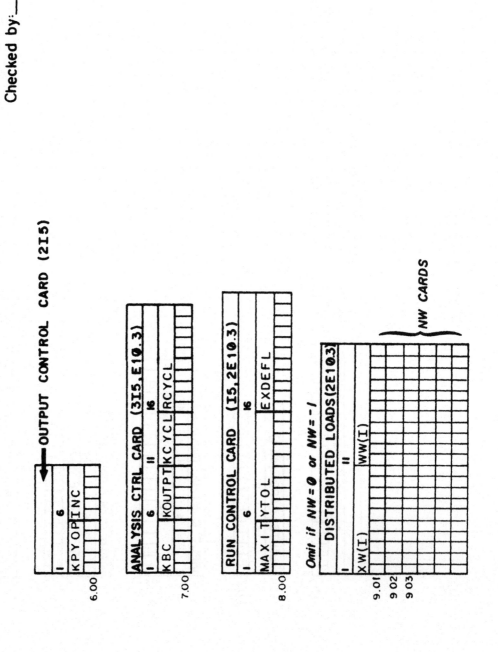

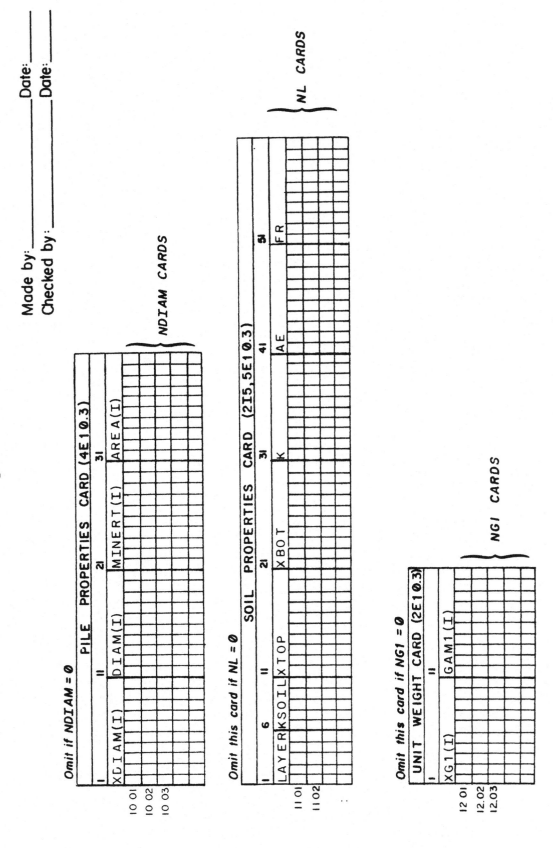

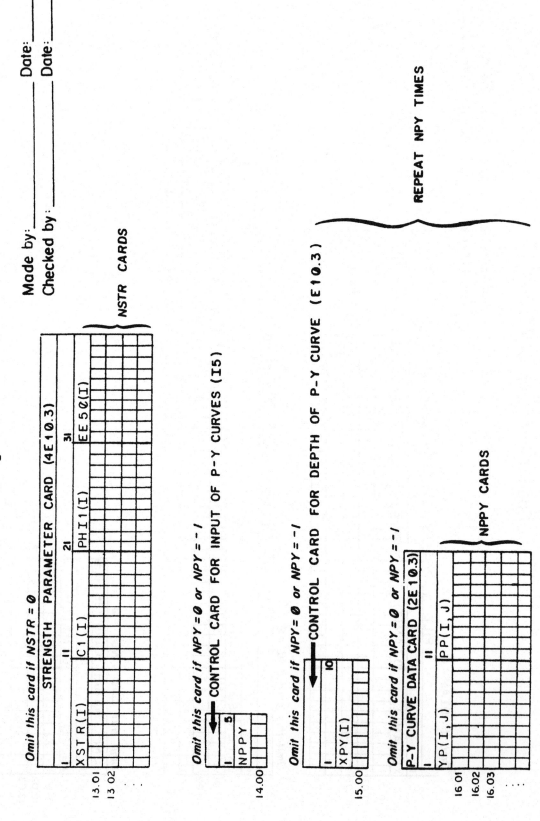

Coding Form for COM624

Made by: _____ Date: _____
Checked by: _____ Date: _____

Omit this card if KPYOP = 0
→ CONTROL CARD FOR OUTPUT OF INTERNALLY-GENERATED P-Y CURVES (I5)

```
| 1    5 |
| NN     |
```
17.00

Omit this card if KPYOP = 0
→ CONTROL CARD FOR DEPTHS AT WHICH INTERNALLY-GENERATED P-Y CURVES ARE TO BE PRINTED (E10.3)

```
| 1        10 |
| XN(I)       |
```
18.01
18.02
⎱ NN CARDS

CARD TO ESTABLISH LOADS ON PILE HEAD (I5, 3E10.3)

```
| 1    6    16    26 |
| KOP  PT   BC2   PX |
```
19.01
19.02
⎱ MAXIMUM 20 LOADINGS

TITLE CARD (18A4)

```
| 1   5   9   13   17   21   25   29   33   37   41   45   49   53   57   61   65   69   72 |
```
20.00

APPENDIX 2

INPUT AND OUTPUT OF COMPUTER PROGRAM COM624 FOR EXAMPLE PROBLEMS

A2.1 INPUT OF EXAMPLE PROBLEM IN SECTION 3.5

Coding Form for COM624

Made by: Shin-Tower Wang   Date: 4/19/84
Checked by: Jim Long   Date: 4/19/84

1.00 TITLE CARD (18A4)
CASE II STIFF CLAY ABOVE WATER TABLE H-PILE STATIC LOADING

2.00 UNITS CARD (4A4)
ENGL

3.00 INPUT CONTROL CARD (4I5)
NI=1, NL=20, NDIAM=1, NW=0

4.00 INPUT CTRL. CARD (3I5)
NGI=2, NSTR=2, NPY=0

5.00 PILE GEOMETRY CARD (3E10.3)
LENGTH=6.00E2, EPILE=2.9@E7, XGS=0.00E0

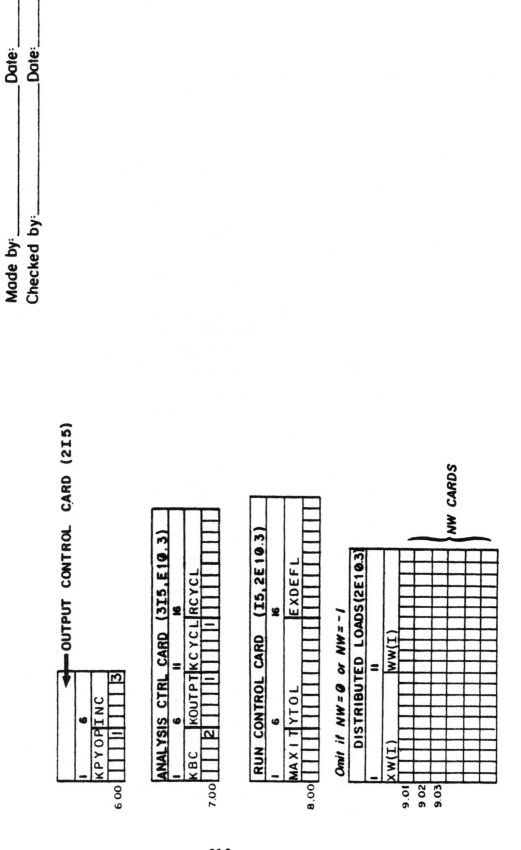

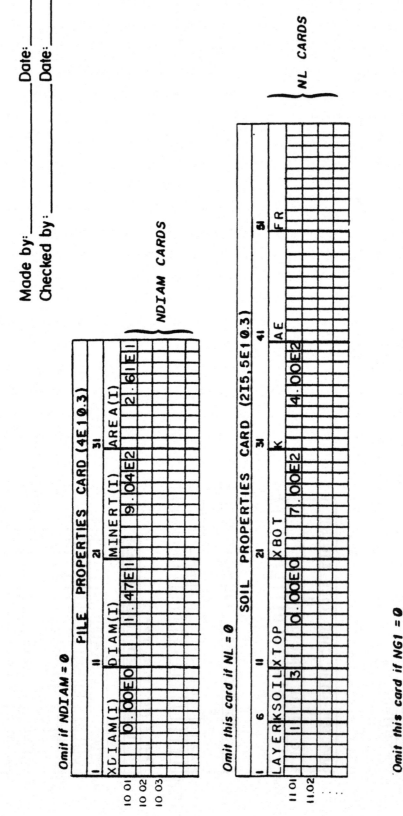

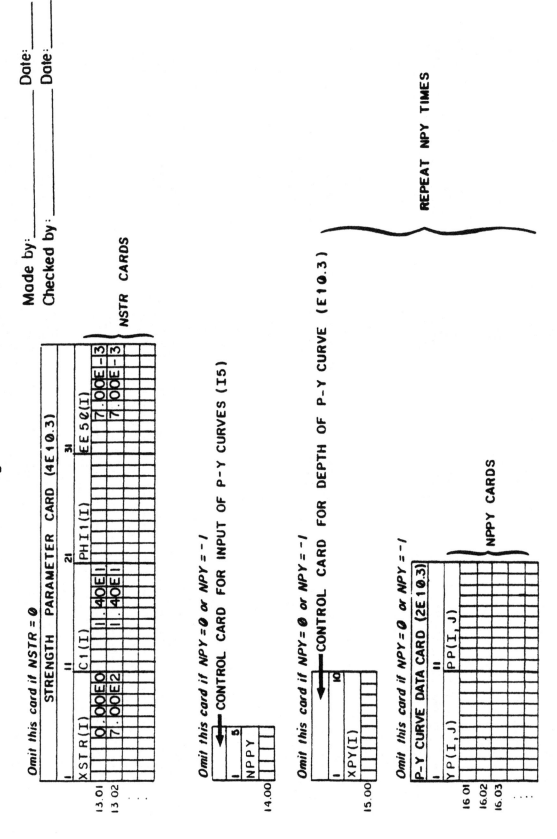

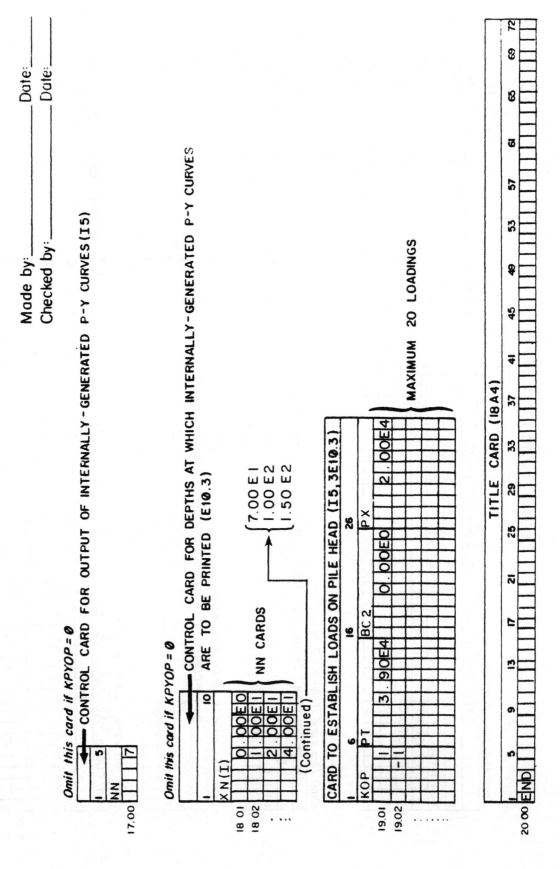

1   CASE1 STIFF CLAY ABOVE WATER TABLE H-PILE STATIC LOADING

UNITS--ENGL

A2.2 OUTPUT OF EXAMPLE PROBLEM IN SECTION 3.5

THE LOADING IS STATIC

PILE GEOMETRY AND PROPERTIES

    PILE LENGTH = 600.00 IN
    MODULUS OF ELASTICITY OF PILE = .290E+08 LBS/IN**2
        1 SECTION(S)

X	DIAMETER	MOMENT OF INERTIA	AREA
IN	IN	IN**4	IN**2
0	14.700	.904E+03	.261E+02
600.00			

SOILS INFORMATION

    X AT THE GROUND SURFACE = 0 IN

    1 LAYER(S) OF SOIL

    LAYER 1
    THE SOIL IS A STIFF CLAY ABOVE THE WATER TABLE
    X AT THE TOP OF THE LAYER = 0 IN
    X AT THE BOTTOM OF THE LAYER = 700.00 IN
    MODULUS OF SUBGRADE REACTION = .400E+03 LBS/IN**3

DISTRIBUTION OF EFFECTIVE UNIT WEIGHT WITH DEPTH
        2 POINTS

X, IN	WEIGHT, LBS/IN**3
0	.69E-01
700.00	.69E-01

DISTRIBUTION OF STRENGTH PARAMETERS WITH DEPTH
        2 POINTS

X, IN	C, LBS/IN**2	PHI, DEGREES	E50
0	.140E+02	0	.700E-02
700.00	.140E+02	0	.700E-02

FINITE DIFFERENCE PARAMETERS
    NUMBER OF PILE INCREMENTS = 120
    TOLERANCE ON DETERMINATION OF DEFLECTIONS = .100E-04 IN

MAXIMUM NUMBER OF ITERATIONS ALLOWED FOR PILE ANALYSIS =   100
MAXIMUM ALLOWABLE DEFLECTION   =   .15E+03 IN

INPUT CODES
    OUTPT = 1
    KCYCL = 1
    KBC   = 2
    KPYOP = 1
    INC   = 3

1   CASE: STIFF CLAY ABOVE WATER TABLE H-PILE STATIC LOADING

UNITS--ENGL

OUTPUT INFORMATION

GENERATED P-Y CURVES

THE NUMBER OF CURVES   = 7
THE NUMBER OF POINTS ON EACH CURVE = 17

DEPTH BELOW GS IN	DIAM IN	C LBS/IN**2	CAVG LBS/IN**2	GAMMA LBS/IN**3	E50
0	14.700	.1E+02	.1E+02	.7E-01	.700E-02

Y IN	P LBS/IN
0	0
.274	313.721
.549	373.379
.823	412.880
1.098	443.669
1.372	469.123
1.646	491.000
1.921	510.291
2.195	527.614
2.470	543.381
2.744	557.884
3.018	571.337
3.293	583.901
3.567	595.703
3.842	606.842
4.116	617.400
5.145	617.400

DEPTH BELOW GS IN	DIAM IN	C LBS/IN**2	CAVG LBS/IN**2	GAMMA LBS/IN**3	E50
10.99	14.700	.1E+02	.1E+02	.7E-01	.700E-02

Y IN	P LBS/IN
0	0

```
                    .274              354.444
                    .549              421.508
                    .823              466.475
                   1.098              501.260
                   1.372              530.015
                   1.646              554.736
                   1.921              576.531
                   2.195              596.192
                   2.470              613.916
                   2.744              630.301
                   3.018              645.500
                   3.293              659.695
                   3.567              673.029
                   3.842              685.615
                   4.116              697.543
                   5.145              697.543
```

```
DEPTH BELOW GS   DIAM        C         CAVG      GAMMA        ESO
     IN          IN      LBS/IN**2  LBS/IN**2  LBS/IN**3
    20.00       14.700    .1E+02     .1E+02      .7E-01      .700E-02
                            Y                     P
                            IN                  LBS/IN
                            0                     0
                           .274              395.168
                           .549              469.936
                           .823              520.470
                          1.098              558.852
                          1.372              590.914
                          1.646              618.471
                          1.921              642.771
                          2.195              664.590
                          2.470              684.451
                          2.744              702.719
                          3.018              719.664
                          3.293              735.490
                          3.567              750.356
                          3.842              764.387
                          4.116              777.686
                          5.145              777.686
```

```
DEPTH BELOW GS   DIAM        C         CAVG      GAMMA        ESO
     IN          IN      LBS/IN**2  LBS/IN**2  LBS/IN**3
    40.00       14.700    .1E+02     .1E+02      .7E-01      .700E-02
                            Y                     P
                            IN                  LBS/IN
                            0                     0
                           .274              476.614
                           .549              566.793
                           .823              627.260
                          1.098              674.034
                          1.372              712.705
                          1.646              745.942
                          1.921              775.250
                          2.195              801.566
                          2.470              825.520
                          2.744              847.553
                          3.018              867.991
                          3.293              887.079
                          3.567              905.009
                          3.842              921.032
```

4.116	937.972
5.145	937.972

DEPTH BELOW GS IN 70.00	DIAM IN 14.700	C LBS/IN**2 .1E+02	CAVG LBS/IN**2 .1E+02	GAMMA LBS/IN**3 .7E-01	E50 .700E-02

Y IN	P LBS/IN
0	0
.274	598.784
.549	712.078
.823	788.044
1.098	846.809
1.372	895.391
1.646	937.148
1.921	973.968
2.195	1007.031
2.470	1037.125
2.744	1064.806
3.018	1090.482
3.293	1114.463
3.567	1136.989
3.842	1158.250
4.116	1178.401
5.145	1178.401

DEPTH BELOW GS IN 100.00	DIAM IN 14.700	C LBS/IN**2 .1E+02	CAVG LBS/IN**2 .1E+02	GAMMA LBS/IN**3 .7E-01	E50 .700E-02

Y IN	P LBS/IN
0	0
.274	720.954
.549	857.364
.823	948.829
1.098	1019.583
1.372	1078.078
1.646	1128.354
1.921	1172.687
2.195	1212.495
2.470	1248.729
2.744	1282.058
3.018	1312.973
3.293	1341.847
3.567	1368.968
3.842	1394.568
4.116	1418.830
5.145	1418.830

DEPTH BELOW GS IN 150.00	DIAM IN 14.700	C LBS/IN**2 .1E+02	CAVG LBS/IN**2 .1E+02	GAMMA LBS/IN**3 .7E-01	E50 .700E-02

Y IN	P LBS/IN
0	0
.274	924.570
.549	1099.506
.823	1216.803
1.098	1307.540

```
                        1.372           1382.555
                        1.646           1447.031
                        1.921           1503.885
                        2.195           1554.036
                        2.470           1601.403
                        2.744           1644.145
                        3.018           1683.791
                        3.293           1720.819
                        3.567           1755.601
                        3.842           1788.430
                        4.116           1819.545
                        5.145           1819.545
```

---------- *** ----------

PILE LOADING CONDITION

    LATERAL LOAD AT PILE HEAD   =   .390E+05 LBS
    SLOPE AT PILE HEAD   =   0   IN/IN
    AXIAL LOAD AT PILE HEAD   =   .200E+05 LBS

X	DEFLECTION	MOMENT	TOTAL STRESS	DISTR. LOAD	SOIL MODULUS	FLEXURAL RIGIDITY
IN	IN	LBS-IN	LBS/IN**2	LBS/IN	LBS/IN**2	LBS-IN**2
0	.139E+00	-.164E+07	.141E+05	0	.191E+04	.262E+11
15.00	.132E+00	-.109E+07	.960E+04	0	.236E+04	.262E+11
30.00	.117E+00	-.603E+06	.567E+04	0	.302E+04	.262E+11
45.00	.956E-01	-.198E+06	.238E+04	0	.399E+04	.262E+11
60.00	.729E-01	.121E+06	.175E+04	0	.550E+04	.262E+11
75.00	.512E-01	.350E+06	.361E+04	0	.795E+04	.262E+11
90.00	.325E-01	.488E+06	.473E+04	0	.123E+05	.262E+11
105.00	.179E-01	.536E+06	.513E+04	0	.210E+05	.262E+11
120.00	.778E-02	.500E+06	.483E+04	0	.423E+05	.262E+11
135.00	.196E-02	.391E+06	.394E+04	0	.128E+06	.262E+11
150.00	-.545E-03	.230E+06	.263E+04	0	.358E+06	.262E+11
165.00	-.106E-02	.930E+05	.153E+04	0	.222E+06	.262E+11
180.00	-.725E-03	.100E+05	.848E+03	0	.294E+06	.262E+11
195.00	-.279E-03	-.262E+05	.979E+03	0	.603E+06	.262E+11
210.00	-.336E-04	-.250E+05	.969E+03	0	.290E+07	.262E+11
225.00	.115E-04	-.519E+04	.800E+03	0	.656E+07	.262E+11
240.00	.151E-05	.172E+04	.780E+03	0	.302E+08	.262E+11
255.00	-.255E-07	-.738E+02	.767E+03	0	.673E+09	.262E+11
270.00	-.141E-16	.183E-03	.766E+03	0	.264E+13	.262E+11
285.00	.440E-22	-.428E-13	.766E+03	0	.264E+13	.262E+11
300.00	-.767E-31	-.732E-18	.766E+03	0	.264E+13	.262E+11
315.00	-.175E-36	.238E-26	.766E+03	0	.264E+13	.262E+11
330.00	.835E-45	.291E-32	.766E+03	0	.264E+13	.262E+11
345.00	.697E-51	-.183E-40	.766E+03	0	.264E+13	.262E+11
360.00	-.543E-59	-.116E-46	.766E+03	0	.264E+13	.262E+11
375.00	-.276E-65	.108E-54	.766E+03	0	.264E+13	.262E+11
390.00	.300E-73	.458E-61	.766E+03	0	.264E+13	.262E+11

405.00	.179E-79	-.568E-69	.766E+03	0	.264E+13 .262E+11
420.00	-.153E-87	-.181E-75	.766E+03	0	.264E+13 .262E+11
435.00	-.031E-94	.281E-83	.766E+03	0	.264E+13 .262E+11
450.00	.730-102	.713E-90	.766E+03	0	.264E+13 .262E+11
465.00	.170-108	-.134E-97	.766E+03	0	.264E+13 .262E+11
480.00	-.346-116	-.280-104	.766E+03	0	.264E+13 .262E+11
495.00	-.667-123	.618-112	.766E+03	0	.264E+13 .262E+11
510.00	.158-130	.110-118	.766E+03	0	.264E+13 .262E+11
525.00	.261-137	-.280-126	.766E+03	0	.264E+13 .262E+11
540.00	-.718-145	-.430-133	.766E+03	0	.264E+13 .262E+11
555.00	-.192-151	.125-140	.766E+03	0	.264E+13 .262E+11
570.00	.314-159	.168-147	.766E+03	0	.264E+13 .262E+11
585.00	.398-166	-.547-155	.766E+03	0	.264E+13 .262E+11
600.00	-.271-173	0	.766E+03	0	.264E+13 .262E+11

OUTPUT VERIFICATION

```
    THE MAXIMUM MOMENT IMBALANCE FOR ANY ELEMENT     = -.285E-05 IN-LBS
    THE MAX. LATERAL FORCE IMBALANCE FOR ANY ELEMENT = -.448E-06 LBS

    COMPUTED LATERAL FORCE AT PILE HEAD     = .39000E+05 LBS
    COMPUTED SLOPE AT PILE HEAD             =      0      IN/IN

    THE OVERALL MOMENT IMBALANCE            = .221E-04 IN-LBS
    THE OVERALL LATERAL FORCE IMBALANCE     = -.105E-05 LBS
```

OUTPUT SUMMARY

```
    PILE HEAD DEFLECTION      =  .139E+00 IN
    MAXIMUM BENDING MOMENT    = -.164E+07 IN-LBS
    MAXIMUM TOTAL STRESS      =  .141E+05 LBS/IN**2
    MAXIMUM SHEAR FORCE       =  .390E+05 LBS

    NO. OF ITERATIONS         =      21
    MAXIMUM DEFLECTION ERROR  =  .626E-05 IN
```

1   CASE: STIFF CLAY ABOVE WATER TABLE H-PILE STATIC LOADING

S U M M A R Y   T A B L E
****

LATERAL LOAD (LBS)	BOUNDARY CONDITION BC2	AXIAL LOAD (LBS)	YT (IN)	ST (IN/IN)	MAX. MOMENT (IN-LBS)	MAX. STRESS (LBS/IN**2)
.390E+05	0	.200E+05	.139E+00	0	-.164E+07	.141E+05

A2.3 Input of example problem in section 3.6

Coding Form for COM624

Made by: Shin-Tower Wang   Date: 4/19/84
Checked by: Jim Long   Date: 4/19/84

TITLE CARD (18A4)

1.00 | CASE2 STIFF CLAY BELOW WATER TABLE H-PILE (CYCLIC LOADING)

UNITS CARD (4A4)

2.00 | ENGL

INPUT CONTROL CARD (4I5)

NI / NL / NDIAM / NW
3.00 | 1 / 20 / / 0

INPUT CTRL. CARD (3I5)

NGI / NSTR / NPY
4.00 | 2 / 2 / 0

PILE GEOMETRY CARD (3E10.3)

LENGTH / EPILE / XGS
5.00 | 6.00E2 / 2.9OE7 / 0.00E0

Coding Form for COM624

Made by: _____ Date: _____
Checked by: _____ Date: _____

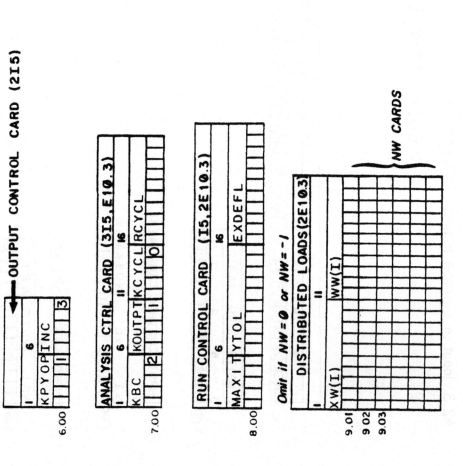

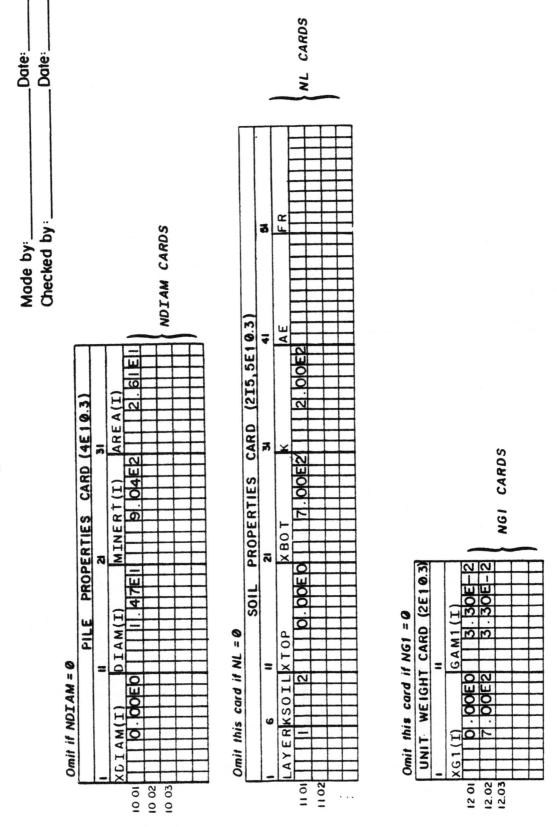

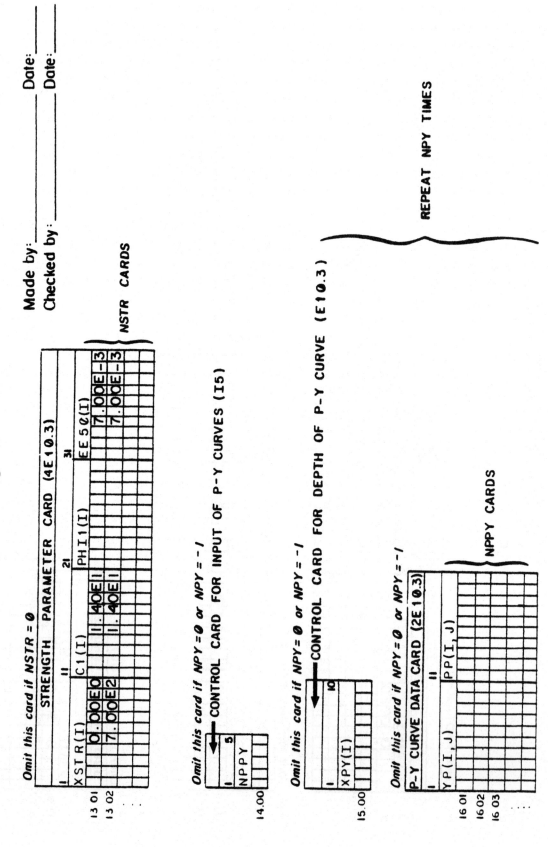

Coding Form for COM624

Made by: _____ Date: _____
Checked by: _____ Date: _____

Omit this card if KPYOP = 0

→ CONTROL CARD FOR OUTPUT OF INTERNALLY-GENERATED P-Y CURVES (I5)

17.00

	5	
NN		

Omit this card if KPYOP = 0

→ CONTROL CARD FOR DEPTHS AT WHICH INTERNALLY-GENERATED P-Y CURVES ARE TO BE PRINTED (E10.3)

	10
XN(I)	

18.01: 1.00E1
18.02
... } NN CARDS

CARD TO ESTABLISH LOADS ON PILE HEAD (I5, 3E10.3)

KOP	6 PT	16 BC2	26 PX
-1	4.00E4	0.00E0	5.00E4

19.01
19.02 .... } MAXIMUM 20 LOADINGS

TITLE CARD (18A4)

20.00: END (columns 5–72)

1   CASE2 STIFF CLAY BELOW WATER TABLE H-PILE (CYCLIC LOADING)

UNITS--ENGL

A2.4 OUTPUT OF EXAMPLE PROBLEM IN SECTION 3.6

```
                    I N P U T   I N F O R M A T I O N
                    ************************************


THE LOADING IS CYCLIC
NO. OF CYCLES =    .12E+03



PILE GEOMETRY AND PROPERTIES
        PILE LENGTH                          =      600.00 IN
        MODULUS OF ELASTICITY OF PILE        =      .290E+08 LBS/IN**2
                  1 SECTION(S)

              X              DIAMETER      MOMENT OF         AREA
                                           INERTIA
             IN                IN           IN**4            IN**2
              0
                              14.700       .904E+03         .201E+02
            600.00



SOILS INFORMATION

      X AT THE GROUND SURFACE            =             0 IN

      1 LAYER(S) OF SOIL

      LAYER  1
      THE SOIL IS A STIFF CLAY BELOW THE WATER TABLE
      X AT THE TOP OF THE LAYER          =             0 IN
      X AT THE BOTTOM OF THE LAYER       =        700.00 IN
      MODULUS OF SUBGRADE REACTION       =      .200E+03 LBS/IN**3

      DISTRIBUTION OF EFFECTIVE UNIT WEIGHT WITH DEPTH
                 2 POINTS
            X,IN       WEIGHT,LBS/IN**3
             0           .33E-01
           700.00        .33E-01

      DISTRIBUTION OF STRENGTH PARAMETERS WITH DEPTH
                 2 POINTS
            X,IN       C,LBS/IN**2      PHI,DEGREES      E50
             0          .140E+02            0            .700E-02
           700.00       .140E+02            0            .700E-02


FINITE DIFFERENCE PARAMETERS
      NUMBER OF PILE INCREMENTS                           =       120
```

```
TOLERANCE ON DETERMINATION OF DEFLECTIONS              =  .100E-04 IN
MAXIMUM NUMBER OF ITERATIONS ALLOWED FOR PILE ANALYSIS =   100
MAXIMUM ALLOWABLE DEFLECTION                           =  .15E+03 IN


INPUT CODES
    OUTPT  =  1
    KCYCL  =  4
    KBC    =  2
    KPYOP  =  1
    INC    =  3
1   CASE2 STIFF CLAY BELOW WATER TABLE H-PILE (CYCLIC LOADING)


UNITS--ENGL
```

OUTPUT INFORMATION
****

GENERATED P-Y CURVES

    THE NUMBER OF CURVES   = 1
    THE NUMBER OF POINTS ON EACH CURVE = 17

DEPTH BELOW GS	DIAM	C	CAVG	GAMMA	E50
IN	IN	LBS/IN**2	LBS/IN**2	LBS/IN**3	
18.00	14.700	.1E+02	.1E+02	.3E-01	.700E-02

    AS =.38   AC =.26

Y,IN	P,LBS/IN
0	0
.008	16.210
.016	32.420
.024	48.630
.032	64.840
.041	81.050
.049	97.260
.057	113.470
.065	129.680
.073	145.890
.081	162.100
.089	178.311
.097	194.521
.162	155.172
.227	111.646
.292	68.119
3.242	68.119

PILE LOADING CONDITION

```
LATERAL LOAD AT PILE HEAD    =   .400E+05 LBS
SLOPE AT PILE HEAD           =    0        IN/IN
AXIAL LOAD AT PILE HEAD      =   .500E+05 LBS
```

X	DEFLECTION	MOMENT	TOTAL STRESS	DISTR. LOAD	SOIL MODULUS	FLEXURAL RIGIDITY
IN	IN	LBS-IN	LBS/IN**2	LBS/IN	LBS/IN**2	LBS-IN**2
0	.120E+00	-.162E+07	.151E+05	0	0	.262E+11
15.00	.113E+00	-.103E+07	.103E+05	0	.247E+04	.262E+11
30.00	.984E-01	-.497E+06	.596E+04	0	.492E+04	.262E+11
45.00	.791E-01	-.754E+05	.253E+04	0	.801E+04	.262E+11
60.00	.599E-01	.209E+06	.362E+04	0	.966E+04	.262E+11
75.00	.406E-01	.367E+06	.490E+04	0	.112E+05	.262E+11
90.00	.253E-01	.422E+06	.535E+04	0	.125E+05	.262E+11
105.00	.136E-01	.406E+06	.521E+04	0	.136E+05	.262E+11
120.00	.529E-02	.348E+06	.474E+04	0	.144E+05	.262E+11
135.00	-.110E-04	.272E+06	.412E+04	0	.149E+05	.262E+11
150.00	-.298E-02	.195E+06	.350E+04	0	.146E+05	.262E+11
165.00	-.427E-02	.128E+06	.295E+04	0	.145E+05	.262E+11
180.00	-.445E-02	.738E+05	.252E+04	0	.145E+05	.262E+11
195.00	-.400E-02	.343E+05	.219E+04	0	.145E+05	.262E+11
210.00	-.324E-02	.787E+04	.108E+04	0	.146E+05	.262E+11
225.00	-.241E-02	-.797E+04	.108E+04	0	.147E+05	.262E+11
240.00	-.164E-02	-.159E+05	.204E+04	0	.147E+05	.262E+11
255.00	-.999E-03	-.183E+05	.206E+04	0	.148E+05	.262E+11
270.00	-.517E-03	-.173E+05	.206E+04	0	.149E+05	.262E+11
285.00	-.182E-03	-.146E+05	.203E+04	0	.149E+05	.262E+11
300.00	.284E-04	-.112E+05	.201E+04	0	.149E+05	.262E+11
315.00	.142E-03	-.792E+04	.198E+04	0	.149E+05	.262E+11
330.00	.188E-03	-.507E+04	.196E+04	0	.149E+05	.262E+11
345.00	.189E-03	-.283E+04	.194E+04	0	.149E+05	.262E+11
360.00	.166E-03	-.123E+04	.193E+04	0	.149E+05	.262E+11
375.00	.132E-03	-.174E+03	.192E+04	0	.149E+05	.262E+11
390.00	.966E-04	.435E+03	.192E+04	0	.149E+05	.262E+11
405.00	.643E-04	.721E+03	.192E+04	0	.149E+05	.262E+11
420.00	.382E-04	.789E+03	.192E+04	0	.149E+05	.262E+11
435.00	.187E-04	.726E+03	.192E+04	0	.149E+05	.262E+11
450.00	.539E-05	.600E+03	.192E+04	0	.149E+05	.262E+11
465.00	-.277E-05	.453E+03	.192E+04	0	.149E+05	.262E+11
480.00	-.743E-05	.315E+03	.192E+04	0	.149E+05	.262E+11
495.00	-.856E-05	.200E+03	.192E+04	0	.149E+05	.262E+11
510.00	-.837E-05	.113E+03	.192E+04	0	.149E+05	.262E+11
525.00	-.718E-05	.534E+02	.192E+04	0	.149E+05	.262E+11
540.00	-.553E-05	.179E+02	.192E+04	0	.149E+05	.262E+11
555.00	-.371E-05	.904E+00	.192E+04	0	.149E+05	.262E+11
570.00	-.187E-05	-.365E+01	.192E+04	0	.149E+05	.262E+11
585.00	-.594E-07	-.193E+01	.192E+04	0	.149E+05	.262E+11
600.00	.173E-05	0	.192E+04	0	.149E+05	.262E+11

OUTPUT VERIFICATION

```
          THE MAXIMUM MOMENT IMBALANCE FOR ANY ELEMENT      =  -.100E+05 IN-LBS
          THE MAX. LATERAL FORCE IMBALANCE FOR ANY ELEMENT =  -.270E-06 LBS

          COMPUTED LATERAL FORCE AT PILE HEAD         =   .40000E+05 LBS
          COMPUTED SLOPE AT PILE HEAD                 =  -.44409E-16 IN/IN

          THE OVERALL MOMENT IMBALANCE                =  -.382E-04 IN-LBS
          THE OVERALL LATERAL FORCE IMBALANCE         =   .124E-05 LBS

  OUTPUT SUMMARY

          PILE HEAD DEFLECTION    =   .120E+00 IN
          MAXIMUM BENDING MOMENT  =  -.162E+07 IN-LBS
          MAXIMUM TOTAL STRESS    =   .151E+05 LBS/IN**2
          MAXIMUM SHEAR FORCE     =   .400E+05 LBS

          NO. OF ITERATIONS       =        13
          MAXIMUM DEFLECTION ERROR =   .809E-05 IN

1    CASE2 STIFF CLAY BELOW WATER TABLE H-PILE (CYCLIC LOADING)


                         S U M M A R Y   T A B L E
                         *****************************
```

LATERAL LOAD (LBS)	BOUNDARY CONDITION BC2	AXIAL LOAD (LBS)	YT (IN)	ST (IN/IN)	MAX. MOMENT (IN-LBS)	MAX. STRESS (LBS/IN**2)
.400E+05	0	.500E+05	.120E+00	-.444E-16	-.162E+07	.151E+05

A2.5 INPUT OF EXAMPLE PROBLEM IN SECTION 3.7 FOR PMEIX

Coding Form for PMEIX

Made by Shin-Tower Wang Date 4/20/84
Checked by Jim Long Date 4/20/84

CARD A (8A10) TITLE DESCRIPTION

1.00 PMEIX FOR CASE 1 (DRILLED SHAFT)

CARD B (2I5) → IDENTIFICATION NUMBER OF THE SHAPE CROSS SECTION AND NUMBER OF LOAD CASES

ISHAPE NP
2.00   10   4

CARD C (F10.2) → AXIAL LOAD

P
3.01   0.00
3.02   200.00   REPEAT NP TIMES
3.03   600.00   (MAXIMUM 10 LOADINGS)
       1200.00

CARD D (4F10.2) STRENGTH AND MODULUS

FC   BARFY   TUBEFY   ES
4.00 4.00   60.00   0.00   29000.00

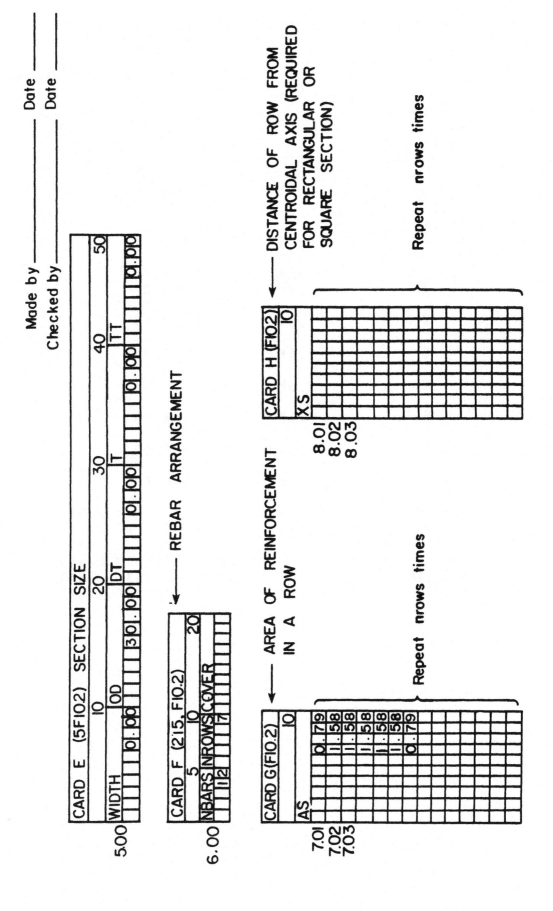

A2.6 OUTPUT OF EXAMPLE PROBLEM IN SECTION 3.7 FOR PMEIX

CIRCULAR SECTION PMEIX

```
SHAPE : CIRCULAR
DIAMETER                    30.00
SHELL THICKNESS                 0
CORE TUBE O.D.                  0
CORE TUBE THICKNESS             0

NO. OF REBARS                  12
ROWS OF REBARS                  7
COVER
(BAR CENTER TO CONCR EDGE)    3.0
```

LAYER	AREA	ORDINATE
1	.79	12.00
2	1.58	10.39
3	1.58	6.00
4	1.58	0
5	1.58	-6.00
6	1.58	-10.39
7	.79	-12.00

```
CONCRETE CYLINDER STRENGTH          4.00KSI
REBARS YIELD STRENGTH              60.00KSI
SHELL/TUBE YIELD STRENGTH              0KSI
MODULUS OF ELAST. OF STEEL      29000.00KSI
MODULUS OF ELAST. OF CONCR       3636.62KSI
SQUASH LOAD CAPACITY             3358.32KPS
```

AXIAL LOAD =   0 KIPS

MOMENT IN KIPS	EI KIP-IN2	PHI	MAX STR IN/IN	N AXIS IN
43.1	43133282.0	.000001	.00001	7.77
215.4	43082893.4	.000005	.00004	7.79
387.3	43031378.1	.000009	.00007	7.80
558.7	42980128.1	.000013	.00010	7.81
759.8	42927717.8	.000017	.00013	7.82
900.4	42875515.6	.000021	.00016	7.83
1070.6	42822118.8	.000025	.00020	7.84
1240.3	42768872.1	.000029	.00023	7.85
1409.6	42715733.8	.000033	.00026	7.87
1578.5	42661336.3	.000037	.00029	7.88
1746.9	42605673.5	.000041	.00032	7.89
1914.8	42550841.1	.000045	.00036	7.90
2082.2	42494396.3	.000049	.00039	7.92
2249.2	42437418.9	.000053	.00042	7.93
3086.3	42004147.3	.000083	.00066	8.01
4664.5	41278986.1	.000113	.00091	8.09
5259.5	36779714.6	.000143	.00113	7.91
5587.9	32300283.8	.000173	.00132	7.65
5775.0	28448323.9	.000203	.00151	7.43
5949.7	25535403.7	.000233	.00169	7.24
6115.0	23250859.9	.000263	.00187	7.11
6142.3	21031741.3	.000293	.00204	6.95
6107.7	19188064.1	.000323	.00219	6.79
6234.0	17608655.0	.000353	.00235	6.66
6259.9	16344384.9	.000383	.00251	6.56
6263.5	15165743.9	.000413	.00268	6.50
6316.3	14258024.3	.000443	.00286	6.45
6339.0	13402457.1	.000473	.00302	6.39
6341.7	12647543.2	.000503	.00319	6.34
6382.0	11975473.1	.000533	.00336	6.30
6403.2	11373383.2	.000563	.00352	6.26
6452.7	10830842.9	.000593	.00370	6.23
6481.9	10340056.5	.000623	.00387	6.21

AXIAL LOAD =   260.00 KIPS

MOMENT IN KIPS	EI KIP-IN2	PHI	MAX STR IN/IN	N AXIS IN
184.7	184738502.8	.000001	.00008	76.53
913.7	182744894.1	.000005	.00014	27.33
1382.6	153627435.0	.000009	.00019	21.09
1686.4	129723879.4	.000013	.00024	18.17
1934.0	113530655.9	.000017	.00028	16.42
2143.0	102049904.7	.000021	.00032	15.21
2341.4	93656511.4	.000025	.00036	14.34
2528.8	87108433.5	.000029	.00040	13.67
2714.2	82126116.7	.000033	.00043	13.14
2887.3	78035979.7	.000037	.00047	12.71
3061.6	74673291.7	.000041	.00051	12.36
3231.6	71812360.4	.000045	.00054	12.06
3401.2	69412568.9	.000049	.00058	11.80
3567.6	67313178.9	.000053	.00061	11.58
4783.5	57632236.5	.000083	.00088	10.58
5954.4	52658533.0	.000113	.00115	10.13
6791.6	47493797.9	.000143	.00140	9.79
7252.8	41923431.3	.000173	.00163	9.45
7474.0	36822411.6	.000203	.00184	9.09
7639.1	32785757.0	.000233	.00206	8.83
7782.1	29589816.0	.000263	.00227	8.62
7944.0	27098817.0	.000293	.00249	8.50
8013.7	24810340.3	.000323	.00270	8.36
8049.4	22774592.4	.000353	.00289	8.19
8063.2	21052661.8	.000383	.00309	8.06
8091.2	19501277.1	.000413	.00330	7.98
8101.5	18207766.4	.000443	.00350	7.89
8109.3	17144439.1	.000473	.00370	7.82
8116.0	16135258.0	.000503	.00390	7.76

AXIAL LOAD =   600.00 KIPS

MOMENT IN KIPS	EI KIP-IN2	PHI	MAX STR IN/IN	N AXIS IN
173.9	172957881.1	.000001	.00021	205.68
864.6	172924483.1	.000005	.00027	53.21
1555.6	172846549.3	.000009	.00033	36.32
2245.7	172743586.7	.000013	.00039	29.87
2865.8	168575406.3	.000017	.00045	26.37
3354.9	159755872.2	.000021	.00050	24.04
3758.3	150331616.3	.000025	.00056	22.35
4099.0	141360008.1	.000029	.00061	21.04
4401.4	133370276.6	.000033	.00066	20.00
4672.7	126290492.1	.000037	.00071	19.14
4924.8	120117225.7	.000041	.00076	18.43
5155.5	114567659.3	.000045	.00080	17.81
5373.8	109670128.2	.000049	.00085	17.28
5581.9	105319732.3	.000053	.00089	16.82
6929.0	83481498.0	.000083	.00121	14.61
8080.8	71511644.4	.000113	.00153	13.51
9119.6	63773183.4	.000143	.00184	12.89
9767.8	56461183.3	.000173	.00215	12.41
10122.1	49862495.4	.000203	.00244	12.01
10394.7	44612652.5	.000233	.00273	11.72
10403.4	39808937.3	.000263	.00302	11.48
10563.0	36051189.0	.000293	.00330	11.27
10616.8	32869264.4	.000323	.00359	11.13
10632.1	30119207.7	.000353	.00390	11.06

AXIAL LOAD =   1200.00 KIPS

MOMENT IN KIPS	EI KIP-IN2	PHI	MAX STR IN/IN	N AXIS IN
153.6	153553530.4	.00001	.00042	418.42
767.6	153515898.3	.00005	.00048	95.77
1380.9	153427813.0	.00009	.00054	59.98
1992.8	153289273.3	.00013	.00060	46.26
2602.7	153100566.2	.00017	.00066	39.02
3210.1	152860877.4	.00021	.00073	34.57
3814.3	152570248.7	.00025	.00079	31.57
4414.3	152216904.1	.00029	.00085	29.41
4970.4	150618946.9	.00033	.00092	27.76
5467.3	147765844.1	.00037	.00098	26.43
5906.3	144055448.9	.00041	.00104	25.33
6302.2	140048137.1	.00045	.00110	24.39
6656.1	135839861.5	.00049	.00116	23.58
6985.6	131804928.9	.00053	.00121	22.89
8803.8	106069602.3	.00083	.00162	19.54
10028.3	88763702.0	.00113	.00202	17.86
10950.9	76369767.4	.00143	.00242	16.91
11572.4	66892245.2	.00173	.00283	16.37
11949.3	58814159.4	.00203	.00327	16.11
12103.4	51902923.4	.00233	.00371	15.93
12110.3	46046738.7	.00263	.00414	15.75

A2.7 INPUT OF EXAMPLE PROBLEM IN SECTION 3.7 FOR COM624

Coding Form for COM624

Made by: Shin Tower Wang   Date: 4/20/84
Checked by: Jim Long   Date: 4/20/84

1.00 TITLE CARD (18A4)
CASE3 STIFF CLAY ABOVE WATER TABLE DRILLED SHAFT (STATIC LOADING)

2.00 UNITS CARD (4A4)
ENGL

3.00 INPUT CONTROL CARD (4I5)
NI=1, NL=20, NDIAM=11, NW=0

4.00 INPUT CTRL. CARD (3I5)
NGI=2, NSTR=2, NPY=0

5.00 PILE GEOMETRY CARD (3E10.3)
LENGTH=6.00E2, EPILE=1.75E6, XGS=0.00E0

Coding Form for COM624

Made by: _____ Date: _____

Checked by: _____ Date: _____

→ OUTPUT CONTROL CARD (2I5)

6.00 | KPYOP | INC | 3 |

ANALYSIS CTRL CARD (3I5, E10.3)

7.00 | KBC | KOUTPT | KCYCL | RCYCL | 2 |

RUN CONTROL CARD (I5, 2E10.3)

8.00 | MAXIT | YTOL | EXDEFL |

Omit if NW = 0 or NW = -1

DISTRIBUTED LOADS (2E10.3)

9.01 | XW(I) | WW(I) |
9.02
9.03

} NW CARDS

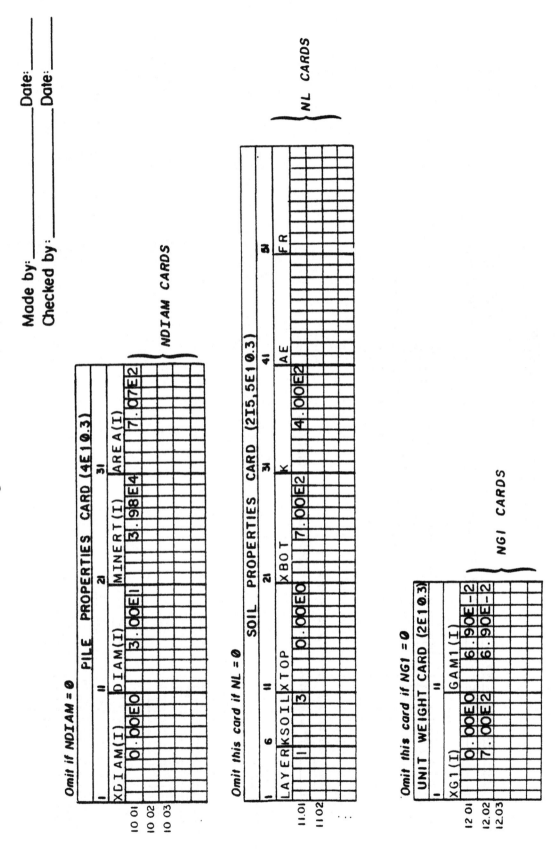

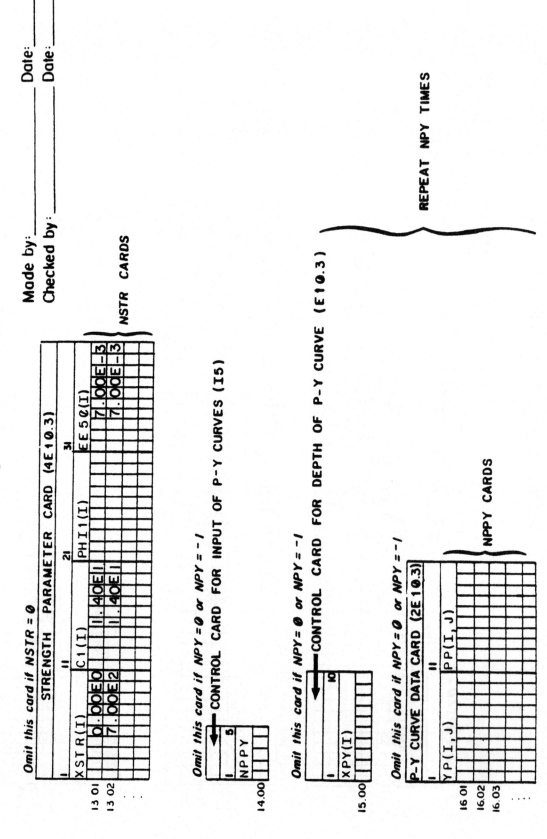

Coding Form for COM624

Made by: _____ Date: _____
Checked by: _____ Date: _____

Omit this card if KPYOP = 0

→ CONTROL CARD FOR OUTPUT OF INTERNALLY-GENERATED P-Y CURVES (I5)

```
         5
|__|NN|_____|
17.00
```

Omit this card if KPYOP = 0

→ CONTROL CARD FOR DEPTHS AT WHICH INTERNALLY-GENERATED P-Y CURVES ARE TO BE PRINTED (E10.3)

```
        10
|__|XN(I)|_____|
18.01  |1.00E1|
18.02
 ...                    } NN CARDS
```

CARD TO ESTABLISH LOADS ON PILE HEAD (I5, 3E10.3)

```
       6         16         26
|__|KOP|  PT  |  BC2  |  PX  |
19.01 |-1|1.16E5|0.00E0|5.00E4|
19.02 ............              } MAXIMUM 20 LOADINGS
```

TITLE CARD (18A4)

```
   5   9  13  17  21  25  29  33  37  41  45  49  53  57  61  65  69  72
|END|
20.00
```

249

1   CASES STIFF CLAY ABOVE WATER TABLE DRILLED SHAFT (STATIC LOADING)

UNITS--ENGL

A2.8 OUTPUT OF EXAMPLE PROBLEM IN SECTION 3.7 FOR COM624

```
             I N P U T   I N F O R M A T I O N
             ***********************************
```

THE LOADING IS STATIC

PILE GEOMETRY AND PROPERTIES

    PILE LENGTH   =   600.00 IN
    MODULUS OF ELASTICITY OF PILE   =   .175E+07 LBS/IN**2
        1 SECTION(S)

            X   DIAMETER   MOMENT OF   AREA
                                       INERTIA
            IN   IN   IN**4   IN**2
            0
                          30.000   .398E+05   .707E+03
        600.00

SOILS INFORMATION

    X AT THE GROUND SURFACE   =   0 IN

    1 LAYER(S) OF SOIL

    LAYER 1
    THE SOIL IS A STIFF CLAY ABOVE THE WATER TABLE
    X AT THE TOP OF THE LAYER   =   0 IN
    X AT THE BOTTOM OF THE LAYER   =   700.00 IN
    MODULUS OF SUBGRADE REACTION   =   .400E+03 LBS/IN**3

    DISTRIBUTION OF EFFECTIVE UNIT WEIGHT WITH DEPTH
                2 POINTS
         X,IN   WEIGHT,LBS/IN**3
           0   .69E-01
         700.00   .69E-01

    DISTRIBUTION OF STRENGTH PARAMETERS WITH DEPTH
                2 POINTS
         X,IN   C,LBS/IN**2   PHI,DEGREES   E50
           0   .140E+02   0   .700E-02
         700.00   .140E+02   0   .700E-02

FINITE DIFFERENCE PARAMETERS
    NUMBER OF PILE INCREMENTS   =   120
    TOLERANCE ON DETERMINATION OF DEFLECTIONS   =   .100E-04 IN

```
MAXIMUM NUMBER OF ITERATIONS ALLOWED FOR PILE ANALYSIS =      100
MAXIMUM ALLOWABLE DEFLECTION                             =    .30E+03 IN


INPUT CODES
    OUTPT  =  1
    KCYCL  =  1
    KBC    =  2
    KPYOP  =  1
    INC    =  3
1   CASE3 STIFF CLAY ABOVE WATER TABLE DRILLED SHAFT (STATIC LOADING)


UNITS--ENGL


           O U T P U T   I N F O R M A T I O N
           *************************************
```

GENERATED P-Y CURVES

    THE NUMBER OF CURVES   = 1
    THE NUMBER OF POINTS ON EACH CURVE   = 17

DEPTH BELOW GS	DIAM	C	CAVG	GAMMA	E50
IN	IN	LBS/IN**2	LBS/IN**2	LBS/IN**3	
12.00	30.000	.1E+02	.1E+02	.7E-01	.70E-02

Y	P
IN	LBS/IN
0	0
.560	680.355
1.120	816.194
1.680	903.268
2.240	970.624
2.800	1026.310
3.360	1074.172
3.920	1116.376
4.480	1154.273
5.040	1188.767
5.600	1220.495
6.160	1249.926
6.720	1277.413
7.280	1303.233
7.840	1327.603
8.400	1350.700
10.500	1350.700

--------- *** ---------

PILE LOADING CONDITION

```
LATERAL LOAD AT PILE HEAD      =   .116E+06 LBS
SLOPE AT PILE HEAD             =   0        IN/IN
AXIAL LOAD AT PILE HEAD        =   .500E+05 LBS
```

X	DEFLECTION	MOMENT	TOTAL STRESS	DISTR. LOAD	SOIL MODULUS	FLEXURAL RIGIDITY
IN	IN	LBS-IN	LBS/IN**2	LBS/IN	LBS/IN**2	LBS-IN**2
0	.438E+00	-.685E+07	.265E+04	0	.138E+04	.697E+11
15.00	.427E+00	-.517E+07	.202E+04	0	.155E+04	.697E+11
30.00	.410E+00	-.365E+07	.145E+04	0	.179E+04	.697E+11
45.00	.362E+00	-.229E+07	.934E+03	0	.210E+04	.697E+11
60.00	.315E+00	-.110E+07	.485E+03	0	.252E+04	.697E+11
75.00	.265E+00	-.847E+05	.103E+03	0	.308E+04	.697E+11
90.00	.215E+00	.745E+06	.351E+03	0	.386E+04	.697E+11
105.00	.167E+00	.139E+07	.594E+03	0	.497E+04	.697E+11
120.00	.124E+00	.184E+07	.765E+03	0	.660E+04	.697E+11
135.00	.866E-01	.211E+07	.868E+03	0	.915E+04	.697E+11
150.00	.559E-01	.221E+07	.903E+03	0	.134E+05	.697E+11
165.00	.323E-01	.213E+07	.875E+03	0	.213E+05	.697E+11
180.00	.156E-01	.190E+07	.788E+03	0	.385E+05	.697E+11
195.00	.496E-02	.154E+07	.651E+03	0	.952E+05	.697E+11
210.00	-.705E-03	.107E+07	.474E+03	0	.428E+06	.697E+11
225.00	-.290E-02	.630E+06	.308E+03	0	.155E+06	.697E+11
240.00	-.304E-02	.288E+06	.179E+03	0	.156E+06	.697E+11
255.00	-.223E-02	.524E+05	.905E+02	0	.205E+06	.697E+11
270.00	-.121E-02	-.812E+05	.101E+03	0	.335E+06	.697E+11
285.00	-.444E-03	-.124E+06	.117E+03	0	.725E+06	.697E+11
300.00	-.574E-04	-.947E+05	.106E+03	0	.337E+07	.697E+11
315.00	.337E-04	-.308E+05	.823E+02	0	.502E+07	.697E+11
330.00	.163E-04	.282E+04	.718E+02	0	.862E+07	.697E+11
345.00	.558E-06	.548E+04	.728E+02	0	.107E+09	.697E+11
360.00	-.541E-07	-.310E+03	.708E+02	0	.636E+09	.697E+11
375.00	.752E-15	.434E+01	.707E+02	0	.922E+13	.697E+11
390.00	.228E-20	-.634E-10	.707E+02	0	.922E+13	.697E+11
405.00	-.533E-29	-.766E-16	.707E+02	0	.922E+13	.697E+11
420.00	-.441E-35	.246E-24	.707E+02	0	.922E+13	.697E+11
435.00	.104E-43	.135E-30	.707E+02	0	.922E+13	.697E+11
450.00	.747E-50	-.672E-39	.707E+02	0	.922E+13	.697E+11
465.00	-.415E-58	-.237E-45	.707E+02	0	.922E+13	.697E+11
480.00	-.124E-64	.161E-53	.707E+02	0	.922E+13	.697E+11
495.00	.951E-73	.417E-60	.707E+02	0	.922E+13	.697E+11
510.00	.218E-79	-.357E-68	.707E+02	0	.922E+13	.697E+11
525.00	-.206E-87	-.731E-75	.707E+02	0	.922E+13	.697E+11
540.00	-.382E-94	.759E-83	.707E+02	0	.922E+13	.697E+11
555.00	.432-102	.128E-89	.707E+02	0	.922E+13	.697E+11
570.00	.667-109	-.157E-97	.707E+02	0	.922E+13	.697E+11
585.00	-.881-117	-.223-104	.707E+02	0	.922E+13	.697E+11
600.00	-.233-123	0	.707E+02	0	.922E+13	.697E+11

OUTPUT VERIFICATION

```
THE MAXIMUM MOMENT IMBALANCE FOR ANY ELEMENT    =   -.239E-04 IN-LBS
```

```
          THE MAX. LATERAL FORCE IMBALANCE FOR ANY ELEMENT =  -.603E-05 LBS

          COMPUTED LATERAL FORCE AT PILE HEAD         =  .11600E+06 LBS
          COMPUTED SLOPE AT PILE HEAD                 =      0      IN/IN

          THE OVERALL MOMENT IMBALANCE                =  -.354E-03 IN-LBS
          THE OVERALL LATERAL FORCE IMBALANCE         =  -.887E-05 LBS

    OUTPUT SUMMARY

          PILE HEAD DEFLECTION       =   .438E+00 IN
          MAXIMUM BENDING MOMENT     =  -.685E+07 IN-LBS
          MAXIMUM TOTAL STRESS       =   .265E+04 LBS/IN**2
          MAXIMUM SHEAR FORCE        =   .116E+06 LBS

          NO. OF ITERATIONS          =       28
          MAXIMUM DEFLECTION ERROR   =   .844E-05 IN
```

1   CASE3 STIFF CLAY ABOVE WATER TABLE DRILLED SHAFT (STATIC LOADING)

                          S U M M A R Y   T A B L E
                          ****

LATERAL LOAD (LBS)	BOUNDARY CONDITION BC2	AXIAL LOAD (LBS)	YT (IN)	ST (IN/IN)	MAX. MOMENT (IN-LBS)	MAX. STRESS (LBS/IN**2)
.116E+06	0	.500E+05	.438E+00	0	-.685E+07	.265E+04

APPENDIX 3

COMPUTER PROGRAM PMEIX

APPENDIX 3

COMPUTER PROGRAM PMEIX

A3.1 INPUT GUIDE

Statement of the Problem

The flexural behavior of a structural element such as a beam, column, or a pile subjected to bending is dependent upon its flexural rigidity which is expressed as the product, EI, of the modulus of elasticity of the material of which it is made and the moment of inertia of the cross section about the axis of bending. In some instances the values of E and I remain constant for all ranges of stresses to which the member is subjected, but there are situations where both E and I vary as the stress conditions change. This variation is most pronounced in reinforced concrete members. Because of nonlinearity in stress-strain relationships, the value of E varies; and because the concrete in the tensile zone below the neutral axis becomes ineffective due to cracking, the value of I is reduced.

When a member is made up of a composite cross section there is no way to calculate directly the value of E for the member as a whole. Reinforced concrete is a composite material, and other examples are concrete encased in a steel tube or a steel section encased in concrete.

Outline of the Solution

The value of EI can be calculated from the moment-curvature relationship of the elastic curve of a beam subjected to bending. Figure A3.1a is a portion of the beam subjected to bending with a radius of curvature ρ. Triangles onn_1 and $n_1 s_1 s_2$ are similar, and

$$\frac{y}{\rho} = \frac{s_1 s_2}{nn_1}, \text{ or}$$

$$\frac{y}{\rho} = \frac{s_1 s_2}{ss_1}, \text{ or}$$

$$\frac{y}{\rho} = \varepsilon, \text{ the strain at the section considered.} \tag{A3.1}$$

and

$$\frac{y}{\rho} = \frac{\sigma}{E},$$

where

$$\sigma = \frac{My}{I}, \text{ and}$$

y = distance of the strained fiber from the neutral axis.

Continuing

$$\frac{y}{\rho} = \frac{1}{E} \times \frac{My}{I}, \text{ and}$$

$$\frac{M}{EI} = \frac{1}{\rho}.$$

From Eq. A3.1

$$\frac{M}{EI} = \frac{\varepsilon}{y}, \text{ and}$$

$$\frac{M}{EI} = \tan \phi, \text{ as is obvious from Fig. A3.1b and c.}$$

Finally,

$$\frac{M}{EI} = \phi, \text{ since } \phi \text{ is very small.} \tag{A3.2}$$

Therefore,

$$EI = \frac{M}{\phi}.$$

Procedure

The procedure consists of calculating the value of M for an assumed value of ϕ and then computing EI from Eq. A3.2. Then a range of values of ϕ, M, and EI can be obtained.

Example

Figure A3.2 shows the cross section of a beam subjected to bending moment. The axial load is 200 kips, $\phi = .0001$ in^{-1}, $E_c = 4000$ kip/sq in, and $E_s = 30,000$ kip/sq in. The values of M and EI are to be found.

As the first step, the position of the neutral axis should be determined by trial, such that the net force on the cross section equals the applied load of 200 kips. Concrete below the neutral axis will be neglected. A linear stress-strain relationship will be assumed here for simplicity.

Trial 1

$$c = 9 \text{ in}$$

Strains:

At top fiber of concrete:	$(.0001)(9) = .0009$
1st row of bars:	$(.0001)(6) = .0006$
2nd row of bars:	$(.0001)(2) = .0002$

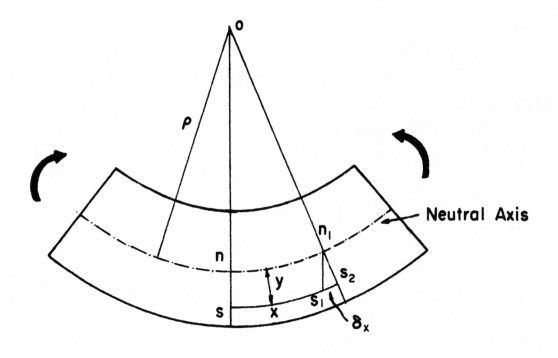

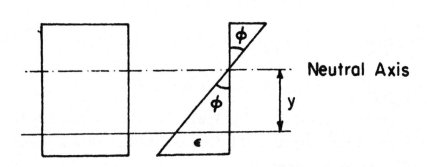

Fig. A3.1. Portion of a beam subjected to bending

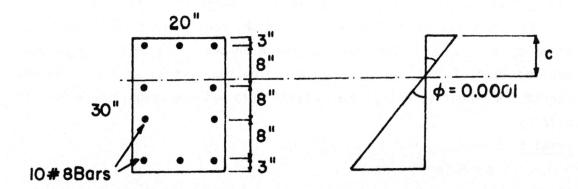

Fig. A3.2. Beam cross-section for example problem

258

3rd row of bars: $(.0001)(10) = .001$
4th row of bars: $(.0001)(18) = .0018$

Forces (stress × area):

Concrete: $\dfrac{(.0009)(4000)}{2}(20)(90) = 324$ kips comp

1st row of bars: $(.0006)(30,000)(3)(.79) = 43$ kips comp

2nd row of bars: $(.0002)(30,000)(2)(.79) = 9$ kips tension

3rd row of bars: $(.001)(30,000)(2)(.79) = 47$ kips tension

4th row of bars: $(.0018)(30,000)(3)(.79) = 128$ kips tension

Net force $= 183$ kips comp

no good

Trial 2

$c = 9.2$ in

Strains:

At top fiber of concrete: $(.0001)(9.2) = .00092$

1st row of bars: $(.0001)(6.2) = .00062$

2nd row of bars: $(.0001)(1.8) = .00018$

3rd row of bars: $(.0001)(9.8) = .00098$

4th row of bars: $(.0001)(17.8) = .00178$

Forces:

Concrete: $(.00092)(\dfrac{4000}{2})(20)(9.2) = 338$ kips comp

1st row of bars: 44 kips comp

2nd row of bars: 8 kips tension

3rd row of bars: 46 kips tension

4th row of bars: 127 kips tension

Net force $= 201$ kips OK

Step 2

The bending moment due to all these forces about the centroidal axis of the cross section now is to be found. Clockwise moments are taken as positive.

Moment due to compression in concrete $= 338\left(15 - \dfrac{(9.2)(1)}{3}\right)$

$= +4033$ in-kips

Moment due to compression in row 1 bars $= (44)(12) = +528$ in-kip

Moment due to tension in row 2 bars $= (8)(4) = -32$ in-kip

Moment due to tension in row 3 bars $= (46)(4) = +1524$ in-kip

Moment due to tension in row 4 bars $= (127)(12) = +1524$ in-kip

Net moment M $= +6237$ in-kip

$EI = \dfrac{M}{\phi} = \dfrac{6237}{.0001} = 63{,}370{,}000$ kips-sq in

The above method, though simple in cases like rectangular cross sections, becomes tedious when cross sections with varying widths are considered. Further, because the actual stress-strain relationship of concrete is a nonlinear function, for a circular cross section the computation of forces will involve double integration, one for area and one for the stress. This is not feasible by hand calculations.

Application to Load-Deflection Analysis of Drilled Shafts or Piles

In the analyses of drilled shafts or piles subjected to bending moments, the flexural rigidity EI is one of the parameters occurring in the differential equation for the solution of deflections. Typically there will be large variations of bending moment along the length of the column. Consequently there will be variations in EI depending on the moment and axial load if any, and this changed value of EI may be employed in calculations.

Program Capabilities

Calculation of forces and moments are done in the program by dividing the cross section into a number of horizontal strips and summing. Figures A3.3 and A3.4 show the stress-strain curves for concrete and steel, respectively, used in the program.

The program gives as output a set of curves for M versus EI values for different axial loads ranging from zero to the axial load capacity for the column. The number of load cases in one run is limited to 10.

Program options allow treatment of the following types of cross sections:

(1) Square or rectangular, reinforced concrete,
(2) Circular, reinforced concrete,
(3) Circular, reinforced concrete, with steel tubular shell around concrete,
(4) Circular, reinforced concrete, with steel tubular shell and tubular core, and
(5) Circular, reinforced concrete, without shell but with tubular core.

Data Input

The data-input form along with the names of variables is shown in Fig. A3.5. The variables are defined in Table A3.1.

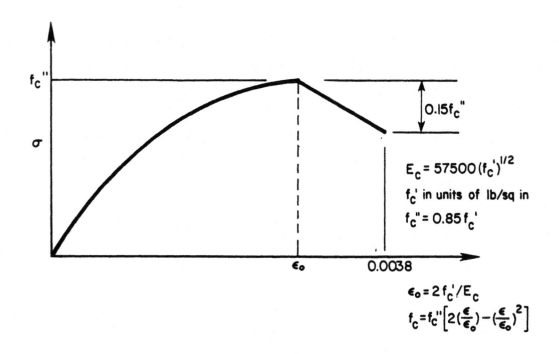

Fig. A3.3. Stress-strain curve for concrete used by program PMEIX

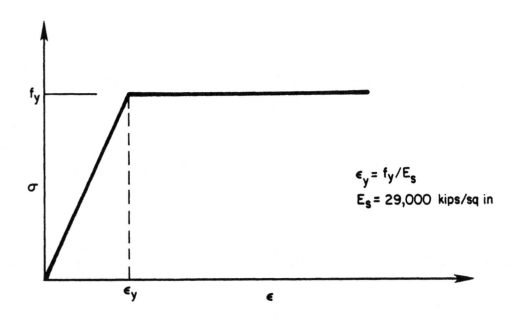

Fig. A3.4. Stress-strain curve for steel used by program PMEIX

Printed Output

The printed output gives a statement of input values, values of E_c and axial load capacity for no moment, and a table each of values of moment, EI, ϕ, maximum strain in the concrete, and depth of neutral axis for each axial load case. The initial value of moment corresponds to a curvature of 0.00001 in^{-1}. The final value is a step higher than the ultimate failure conditions of concrete. Values at failure can be read at a concrete strain of 0.0030.

Other Output

If an axial load more than the collapse load is applied, the program stops and an error message is printed out.

All units in inches and kips

```
         10        20        30        40        50        60        70        80
A                ANAME
B       ISHAPE    5        10
C          P    10            (Repeat NP times)
D          FC   10    BARFY  20   TUBEFY  30   ES  40
E       WIDTH   10    OD     20   DT      30   T   40   TT  50
F       NBARS    5    NROWS  10   COVER   20
G          AS   10            (Repeat NROWS times)
H          XS   10            (Repeat NROWS times)    Card H required only if ASHAPE is 1, i.e., for
                                                      rectangular or square cross sections.
```

Fig. A3.5. Data input form for Computer Program PMEIX. A description of variables is given in Table A3.1.

TABLE A3.1. DETAILED INPUT GUIDE WITH DEFINITIONS OF VARIABLES

(All units in inches and kips)

Card A

(8A10)

ANAME	Alphanumeric description to be printed as title

Card B

(2I5)

ISHAPE	Identification number of the shape of cross section of column/pile 1: Rectangular or square 10: Circular (without shell or core) 20: Circular (with shell but without core) 30: Circular (with shell and core or without shell and with core)
NP	Number of load cases (axial)

Card C

(F10.2)

P	Axial load. The total number of axial loads per run is limited to 10.

Card D

(4F10.2)

FC	Cylinder strength of concrete
BARFY	Yield strength of reinforcement
TUBEFY	Yield strength of shell or core
ES	Modulus of elasticity of steel

Card E

(5F10.2)

WIDTH	Width of section if rectangular (0.0 if circular)
OD	Outer diameter, if circular, or depth of section if rectangular
DT	Outer diameter of core (0.0 if ISHAPE is 1 or 10)
T	Thickness of shell
TT	Thickness of core

(TABLE Continued)

TABLE A3.1. DETAILED INPUT GUIDE WITH DEFINITIONS OF VARIABLES
(cont.)

Card F

(2I5,F10.2)

NBARS	Number of reinforcing bars
NROWS	Number of rows of reinforcing bars (a number not exceeding 50)
COVER	Cover of rebar, from center of rebar to outer edge of concrete

Card G

(F10.2)

AS   Area of reinforcement in a row
AS (1) is for the top row
AS (2) is for the 2nd row from the top, etc.
The total number of values should not exceed 50

Note: In the cases of an odd number of bars in a circular cross section the centroidal axis is taken as the diameter passing through one bar. In this case the number of rows will be the same as the number of bars.

Card H

(F10.2)

XS   Distance of row from centroidal axis, starting from top row downwards. Positive for rows above the axis and negative for rows below the axis. The total number of values should not exceed 50.

Note: Card H is required only in the case of rectangular or square sections.

A3.2 EXAMPLE PROBLEMS

Two example problems are solved. Figures A3.6a and A3.6b give the cross sections of each example. For each problem three load cases, 0.0, 10.0, and 1000 kips are used. The listing of the input for the example problems and the output are shown in the following sections.

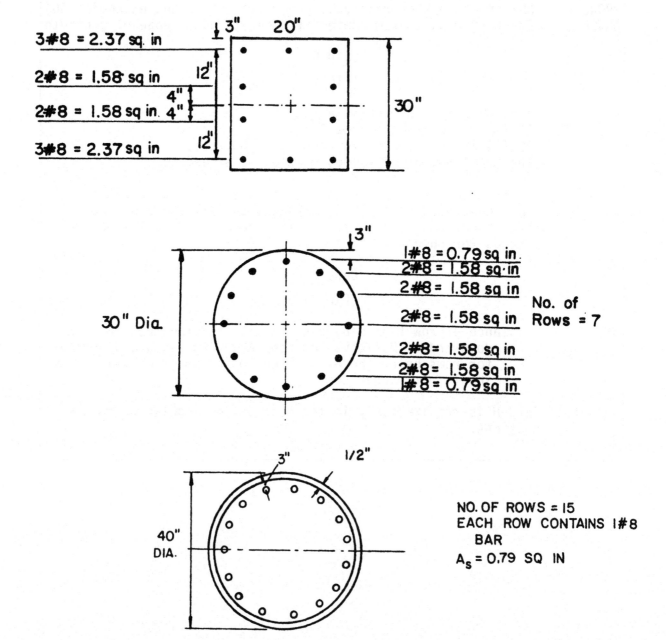

Fig. A3.6. Concrete column cross-sections for example problem

A3.3 LISTING OF PROGRAM

```
      PROGRAM PMEIX (INPUT,OUTPUT)
C
C     THIS PROGRAM GIVES A SET OF VALUES OF -EI- FOR VARIOUS VALUES
C     OF MOMENTS COMBINED WITH AXIAL LOADS RANGING FROM ZERO TO ANY
C     LOAD LESS THAN THE SQUASH LOAD.
C
C     THE PROGRAM CAN TREAT SQUARE, RECTANGULAR OR CIRCULAR SHAPES
C     OF CONCRETE WITH REINFORCEMENT OF ANY GRADE.  CIRCULAR SHAPES
C     CAN BE SPECIFIED AS WITH OR WITHOUT A STEEL SHELL, WITH OR
C     WITHOUT A TUBULAR STEEL CORE (NO CONCRETE IN CORE).
C
C     ---- ALL UNITS SHALL BE INPUT IN INCHES AND KIPS ----
C
C     INPUT FORMATS ARE AS FOLLOWS:
C
C     1. FORMAT(8A10). THIS LINE IS FOR IDENTIFICATION OF THE
C     PROBLEM.  WILL READ ANY CHARACTER IN THE FIRST 80 COLUMNS.
C
C     2. FORMAT(2I5). THIS LINE IS FOR SHAPE IDENTIFICATION AND
C     NUMBER OF LOAD CASES.
C           SHAPE IDENTIFICATION:
C              1=RECTANGULAR
C             10=CIRCULAR WITHOUT SHELL OR CORE
C             20=CIRCULAR WITH SHELL, WITHOUT CORE
C             30=CIRCULAR WITH SHELL AND CORE
C     THE NUMBER OF LOAD CASES SHALL NOT EXCEED TEN (10).
C
C     3. FORMAT(F10.2). THIS LINE IS FOR THE APPLIED AXIAL LOADS.
C     ONE (1) LINE FOR EACH LOAD CASE.
C
C     4. FORMAT(4F10.2). THIS LINE IS FOR THE COMPRESSIVE STRENGTH
C     OF THE CONCRETE, YIELD STRENGTH OF THE REINFORCEMENT, YIELD
C     STRENGTH OF THE SHELL OR CORE STEEL, AND THE MODULUS OF
C     ELASTICITY OF THE STEEL.
C
C     5. FORMAT(5F10.2). THIS LINE IS FOR THE WIDTH OF THE SECTION
C     (SPECIFY AS 0.0 IF CIRCULAR), DEPTH OF THE SECTION (EXTERNAL
C     DIAMETER IF CIRCULAR), EXTERNAL DIAMETER OF THE INNER TUBE,
C     THICKNESS OF THE OUTER SHELL, AND THE THICKNESS OF THE INNER
C     TUBE
C
C     6. FORMAT(2I5,F10.2). THIS LINE IS FOR THE NUMBER OF REBARS,
C     THE NUMBER OF ROWS OF REBARS, AND THE CONCRETE COVER FROM THE
C     CENTER OF THE REBAR TO THE EDGE OF THE CONCRETE.
C
C     7. FORMAT(F10.2)  THIS LINE IS FOR THE AREA OF REINFORCEMENT
C     IN EACH ROW, STARTING FROM THE TOP ROW OF THE SECTION.
C     ONE LINE FOR EACH ROW OF REBAR.  THE NUMBER OF ROWS SHALL
C     NOT EXCEED FIFTY (50).
C
C     8. FORMAT(F10.2). THIS LINE IS REQUIRED ONLY IF THE SECTION
C     IS RECTANGULAR OR SQUARE.  IT IS FOR THE DISTANCE FROM THE
C     CENTROIDAL AXIS TO EACH ROW OF REBAR, STARTING FROM THE TOP
C     ROW.  VALUES ARE POSITIVE (+) IF THE ROW IS ABOVE THE AXIS,
C     AND NEGATIVE (-) IF THE ROW IS BELOW THE AXIS.
C     ONE LINE FOR EACH ROW.
C
C
      COMMON/ONE/PCON,AMCON,PSTEEL,FFC,PSHELL,AMSHELL,AC,PSHT,LL(50)
      COMMON/TWO/PCORE,AMCORE,ISHAPE,WIDTH,OD,RO,RC,DC,DT,T,TT,Z,COVER
      COMMON/THREE/NROWS,NP,FC,K,DEL,PHI,RINT,RINTS,DNA,C,F,G,TUBEFY
```

```
      COMMON/FOUR/KA,II,KK,PCAP,RCS,ROS,RT,RTS,DIF,ASH(100),IJ,PCT,NBARS
      COMMON/FIVE/P(10),AS(50),XS(50),SSN(50),SSS(50),AXS(50),BARFY,STR
      COMMON/SIX/X(100),CSN(100),CSS(100),ACC(100),ATOT(100),SSM(100)
      COMMON/SEVEN/RSSM(100),XX(100),RXX(100),AM(50),XXS(100),RXXS(100)
      COMMON/EIGHT/TA(100),RTA(100),TACC(100),RACC(100),RX(100),ES,EC,IK
      COMMON/NINE/RASH(100),RATOT(100),RTOT(100),RTACC(100),TOT(100)
      DIMENSION ANAME(8)
      DIMENSION DDNA(2)
      REAL NDNA
      READ 20,(ANAME(I),I=1,8)
      READ 1,ISHAPE,NP
      READ 2,(P(J),J=1,NP)
      READ 3,FC,BARFY,TUBEFY,ES
      READ 4,WIDTH,OD,DT,T,TT
      PRINT 22,(ANAME(I),I=1,8)
      IF (ISHAPE.EQ.1) GOTO 18
      PRINT 6,OD,T,DT,TT
      GOTO 19
   18 PRINT 7,WIDTH,OD
   19 CALL SETUP
      PRINT 5,NBARS
      PRINT 8,NROWS
      PRINT 11,COVER
      PRINT 9
      PRINT 10,(JJ,AS(JJ),XS(JJ),JJ=1,NROWS)
      PRINT 12,FC
      PRINT 13,BARFY
      PRINT 14,TUBEFY
      PRINT 15,ES
      PRINT 16,EC
      PRINT 17,PCAP
      DO 500 J=1,NP
      IF (PCAP-P(J)) 430,26,26
   26 PRINT 30,P(J)
      PRINT 21
      PRINT 23
      PHI=.000001
   35 NSWTCH=0
      DDNA(1)=0.0
      DDNA(2)=OD
      DNA=OD
   40 CALL CSTRESS
      PCON=0.0
      IF (DNA-OD) 33,33,55
   33 K=INT(DNA/DEL)
      GO TO 481
   55 K=60
  481 CONTINUE
      DO 90 I=1,K
      XX(I)=X(I)+RO-DNA
      IF (ISHAPE.EQ.1) GOTO 85
      XXS(I)=XX(I)*XX(I)
      IF (RINT.GE.ABS(XX(I))) GOTO 47
      ACC(I)=0.0
      GOTO 44
   47 ACC(I)=2.*SQRT(RINTS-XXS(I))*DEL
   44 PCON=PCON+CSS(I)*ACC(I)
      GOTO 90
   85 PCON=PCON+AC*CSS(I)
   90 CONTINUE
      CALL STEELP
      PTOT=PCON+PSTEEL
```

```
      IF (ISHAPE.EQ.1) GOTO 86
      IF (ISHAPE.EQ.10) GOTO 86
      CALL SHELLP
      PTOT=PTOT+PSHELL
      IF (ISHAPE.EQ.20) GOTO 86
      CALL COREP
      PTOT=PTOT+PCORE
86    IF(PTOT .GT. P(J)) GOTO 89
      DDNA(1)=DNA
      IF(NSWTCH .EQ. 1)GOTO 92
      DDNA(2)=5.0*DNA
      GOTO 92
89    DDNA(2)=DNA
      NSWTCH=1
92    NDNA=(DDNA(1)+DDNA(2))/2.0
      IF(ABS(DNA-NDNA) .LE. .0001) GOTO 50
      DNA=NDNA
      GOTO 40
50    AMCON=0.0
      DO 100 I=1,K
      IF(ISHAPE.EQ.1) GOTO 65
      AMCON=AMCON+CSS(I)*ACC(I)*(XX(I))
      GOTO 100
65    AMCON=AMCON+AC*CSS(I)*(XX(I))
100   CONTINUE
      AMSTEEL=0.0
      DO 200 I=1,NROWS
200   AMSTEEL=AMSTEEL+SSS(I)*AS(I)*XS(I)
      AMTOT=AMCON+AMSTEEL
      IF (ISHAPE.EQ.1) GOTO 110
      IF (ISHAPE.EQ.10) GOTO 110
      CALL SHELLM
      AMTOT=AMTOT+AMSHELL
      IF (ISHAPE.EQ.20) GOTO 110
      CALL COREM
      AMTOT=AMTOT+AMCORE
110   EI=AMTOT/PHI
      CSNMAX=PHI*(DNA-T)
      PRINT 400,AMTOT,EI,PHI,CSNMAX,DNA
      IF (CSNMAX.GT..0038) GOTO 500
      GOTO 450
430   PRINT 400
      GOTO 500
450   IF (PHI.GT..00005) GOTO 460
      PHI=PHI+.000004
      GOTO 35
460   PHI=PHI+.00003
      GOTO 35
500   CONTINUE
1     FORMAT (2I5)
2     FORMAT (F10.2)
3     FORMAT (4F10.2)
4     FORMAT (5F10.2)
5     FORMAT(5X,*NO. OF REBARS             *,I5)
6     FORMAT (5X,*SHAPE : CIRCULAR*/,5X,*DIAMETER  *,10X,F5.2,/,5X,
     **SHELL THICKNESS   *,F5.2,/,5X,*CORE TUBE O.D.      *,
     *F5.2/,5X,*CORE TUBE THICKNESS *,F5.2/)
7     FORMAT (5X,*SHAPE : RECTANGULAR*,/,5X,*WIDTH             *,
     *F5.2,5X,*DEPTH           *,F5.2/)
8     FORMAT (5X,*ROWS OF REBARS           *,I5)
9     FORMAT(5X,*LAYER*,11X,*AREA*,7X,*ORDINATE*)
10    FORMAT(5X,I5,5X,F10.2,5X,F10.2/)
```

```
11    FORMAT (5X,*COVER*,/,5X,*(BAR CENTER TO CONCR EDGE)*,
     *3X,F3.1/)
12       FORMAT (5X,*CONCRETE CYLINDER STRENGTH    *,F10.2,*KSI*)
13    FORMAT (5X,*REBARS YIELD STRENGTH        *,F10.2,*KSI*)
14    FORMAT (5X,*SHELL/TUBE YIELD STRENGTH*,5X,F10.2,*KSI*)
15    FORMAT (5X,*MODULUS OF ELAST. OF STEEL    *,F10.2,*KSI*)
16    FORMAT (5X,*MODULUS OF ELAST. OF CONCR    *,F10.2,*KSI*)
17    FORMAT (5X,*SQUASH LOAD CAPACITY        *,F10.2,*KPS*//)
20    FORMAT(8A10)
22    FORMAT(*1*,32X,8A10)
30    FORMAT(*1*,4X,*AXIAL LOAD = *,F10.2,2X,*KIPS*)
21    FORMAT(/13X,*MOMENT        EI        PHI    MAX STR    N AXIS*)
23    FORMAT(13X,*IN KIPS     KIP-IN2            IN/IN       IN*)
400   FORMAT(10X,F9.1,2X,F12.1,2X,F7.6,2X,F7.5,2X,F6.2)
440   FORMAT(5X,*APPLIED LOAD MORE THAN SQUASH LOAD*)
      STOP
      END
      SUBROUTINE SETUP
      COMMON/ONE/PCON,AMCON,PSTEEL,FFC,PSHELL,AMSHELL,AC,PSHT,LL(50)
      COMMON/TWO/PCORE,AMCORE,ISHAPE,WIDTH,OD,RO,RC,DC,DT,T,TT,Z,COVER
      COMMON/THREE/NROWS,NP,FC,K,DEL,PHI,RINT,PINTS,DNA,C,F,G,TUBEFY
      COMMON/FOUR/KA,II,KK,PCAP,RCS,ROS,RT,RTS,DIF,ASH(100),IJ,PCT,NBARS
      COMMON/FIVE/P(10),AS(50),XS(50),SSN(50),SSS(50),AXS(50),BARFY,STR
      COMMON/SIX/X(100),CSN(100),CSS(100),ACC(100),ATOT(100),SSH(100)

      COMMON/SEVEN/RSSH(100),XX(100),RXX(100),AM(50),XXS(100),RXXS(100)
      COMMON/EIGHT/TA(100),RTA(100),TACC(100),RACC(100),RX(100),FS,EC,IK
      COMMON/NINE/RASH(100),RATOT(100),RTOT(100),RTACC(100),TOT(100)
      FACT=.7854
      EC=57.5*SQRT(FC*1000.)
      STP=1.7*FC/EC
      FFC=FC
      DEL=OD/60.
      RO=.5*OD
      READ 1,NBARS,NROWS,COVER
      READ 2,(AS(JJ),JJ=1,NROWS)
      IF(ISHAPE.EQ.1) GOTO 300
      ANBARS=NBARS
      LA=INT(ANBARS/2.)
      LB=2*LA
      NA=INT(ANBARS/4.)
      NB=4*NA
      ANGLE=6.2832/ANBARS
      RS=RO-COVER-T
      DO 200 JJ=1,NROWS
      AJJ=JJ
      IF ( ANBARS-LB ) 7,6,7
7     AM(JJ)=(ANBARS+1.)/4.-AJJ/2.
      GOTO 200
6     IF(NBARS.EQ.NB) GOTO 100
      AM(JJ)=ANBARS/4.-AJJ+.5
      GOTO 200
100   AM(JJ)=ANBARS/4.-AJJ+1.
200   XS(JJ)=RS*SIN(ANGLE*AM(JJ))
      GOTO 4
300   READ 3,(XS(JJ),JJ=1,NROWS)
4     ARS=0.0
      DO 5 I=1,NROWS
5     ARS=ARS+AS(I)
      IF(ISHAPE.EQ.1)GOTO 101
      D=OD-T-T
      ROS=RO*RO
```

```
      RINT=.5*D
      RINTS=RINT*RINT
      ARC=FACT*D*D-ARS
      A=FACT*((OD**2)-(D**2))
      IF(ISHAPE.EQ.10) GOTO 103
      IF(ISHAPE .EQ. 20) GOTO 103
      RT=.5*DT
      DC=DT-TT-TT
      RC=.5*DC
      RTS=RT*RT
      RCS=RC*RC
      Z=RO+RT
      ATUBE=FACT*((DT**2)-(DC**2))
      AIN=FACT*DT*DT
      PCAP=FFC*(ARC-AIN)+BARFY*ARS+TUBEFY*(A+ATUBE)
      DIF=RO-RT
      GOTO 104
  103 PCAP=FFC*ARC+BARFY*ARS+TUBEFY*A
      GOTO 104
  101 AC=WIDTH*DEL
      ARC=WIDTH*OD-ARS
      PCAP=FFC*ARC+BARFY*ARS
   1  FORMAT(2I5,F10.2)
   2  FORMAT(F10.2)
   3  FORMAT(F10.2)
  104 RETURN
      END
      SUBROUTINE CSTRESS
      COMMON/ONE/PCON,AMCON,PSTEEL,FFC,PSHELL,AMSHELL,AC,PSHT,LL(50)
      COMMON/TWO/PCORE,AMCORE,ISHAPE,WIDTH,OD,RO,RC,DC,DT,T,TT,Z,COVER
      COMMON/THREE/NROWS,NP,FC,K,DEL,PHI,RINT,RINTS,DNA,C,F,G,TUBEFY
      COMMON/FIVE/P(10),AS(50),XS(50),SSN(50),SSS(50),AXS(50),BARFY,STR
      COMMON/SIX/X(100),CSN(100),CSS(100),ACC(104),ATOT(100),SSH(100)
      N=INT(DNA/DEL)
      C=DNA-(FLOAT(N))*DEL
      F=DEL-C
      K=60
      IF(DNA-OD) 10,10,20
  10  X(1)=C+.5*DEL
      GOTO 60
  20  X(1)=(DNA-OD)+.5*DEL
  60  DO 90 I=1,K
      IF (I.GT.1) X(I)=X(1)+(FLOAT(I-1))*DEL
      CSN(I)=X(I)*PHI
      IF (CSN(I)-STR) 70,70,80
  70  CSS(I)=FFC*(2.*CSN(I)/STR-(CSN(I)/STR)**2)
      GOTO 90
  80  CSS(I)=FFC*(.85+.15*(.0038-CSN(I))/(.0038-STR))
  90  CONTINUE
      RETURN
      END
      SUBROUTINE STEELP
      COMMON/ONE/PCON,AMCON,PSTEEL,FFC,PSHELL,AMSHELL,AC,PSHT,LL(50)
      COMMON/TWO/PCORE,AMCORE,ISHAPE,WIDTH,OD,RO,RC,DC,DT,T,TT,Z,COVER
      COMMON/THREE/NROWS,NP,FC,K,DEL,PHI,RINT,RINTS,DNA,C,F,G,TUBEFY
      COMMON/FIVE/P(10),AS(50),XS(50),SSN(50),SSS(50),AXS(50),BARFY,STR
      COMMON/EIGHT/TA(100),RTA(100),TACC(100),RACC(100),RX(100),ES,EC,IK
      PSTEEL=0.0
      DO 100 JJ=1,NROWS
      AXS(JJ)=XS(JJ)+DNA-RO
      SSN(JJ)=PHI*AXS(JJ)
      SSS(JJ)=SSN(JJ)*ES
```

```
      IF (SSS(JJ).GT.BARFY) SSS(JJ)=BARFY
      IF (SSS(JJ).LT.-BARFY) SSS(JJ)=-BARFY
      PSTEEL=PSTEEL+AS(JJ)*SSS(JJ)
  100 CONTINUE
      RETURN
      END
      SUBROUTINE SHELLP
      COMMON/ONE/PCON,AMCON,PSTEEL,FFC,PSHELL,AMSHELL,AC,PSHT,LL(50)
      COMMON/TWO/PCORE,AMCORE,ISHAPE,WIDTH,OD,RO,RC,DC,DT,T,TT,Z,COVER
      COMMON/THREE/NROWS,NP,FC,K,DEL,PHI,RINT,RINTS,DNA,C,F,G,TUBEFY
      COMMON/FOUR/KA,II,KK,PCAP,RCS,ROS,RT,RTS,DIF,ASH(100),TJ,PCT,NBARS
      COMMON/FIVE/P(10),AS(50),XS(50),SSN(50),SSS(50),AXS(50),BARFY,STR
      COMMON/SIX/X(100),CSN(100),CSS(100),ACC(100),ATOT(100),SSH(100)
      COMMON/SEVEN/RSSH(100),XX(100),RXX(100),AM(50),XXS(100),RXXS(100)
      COMMON/EIGHT/TA(100),RTA(100),TACC(100),RACC(100),RX(100),ES,EC,IK
      COMMON/NINE/RASH(100),RATOT(100),RTOT(100),RTACC(100),TOT(100)
      DIMENSION SM(100)
      PSHT=0.0
      PSHELL=0.0
      DO 103 I=1,K
      IF (RO.GT.ABS(XX(I))) GOTO 110
      ASH(I)=0.0
      GOTO 106
  110 ATOT(I)=2.*SQRT(ROS-XXS(I))*DEL
      ASH(I)=ATOT(I)-ACC(I)
  106 SSH(I)=ES*CSN(I)
      IF (SSH(I).GT.TUBEFY) SSH(I)=TUBEFY
      IF (SSH(I).LT.-TUBEFY) SSH(I)=-TUBEFY
  103 PSHELL=PSHELL+ASH(I)*SSH(I)
      IF (DNA-OD) 107,105,105
  107 KA=99-K
      DO 104 I=1,KA
      RX(I)=X(I)-C+F
      RXX(I)=RO-DNA-RX(I)
      RXXS(I)=RXX(I)*RXX(I)
      IF (RINT.GT.ABS(RXX(I))) GOTO 47
      RACC(I)=0.0
      GOTO 44
   47 RACC(I)=2.*SQRT(RINTS-RXXS(I))*DEL
   44 IF (RO.GT.ABS(RXX(I))) GOTO 48
      RASH(I)=0.0
      GOTO 104
   48 RATOT(I)=2.*SQRT(ROS-RXXS(I))*DEL
      RASH(I)=RATOT(I)-RACC(I)
      SM(I)=RX(I)*PHI
      RSSH(I)=ES*SM(I)
      IF (RSSH(I).GT.TUBEFY) RSSH(I)=TUBEFY
      IF (RSSH(I).LT.-TUBEFY) RSSH(I)=-TUBEFY
  104 PSHT=PSHT+RASH(I)*RSSH(I)
      G=F*PHI*ES*T*DEL
      PSHELL=PSHELL-PSHT-G
  105 RETURN
      END
      SUBROUTINE SHELLM
      COMMON/ONE/PCON,AMCON,PSTEEL,FFC,PSHELL,AMSHELL,AC,PSHT,LL(50)
      COMMON/TWO/PCORE,AMCORE,ISHAPE,WIDTH,OD,RO,RC,DC,DT,T,TT,Z,COVER
      COMMON/THREE/NROWS,NP,FC,K,DEL,PHI,RINT,RINTS,DNA,C,F,G,TUBEFY
      COMMON/FOUR/KA,II,KK,PCAP,RCS,ROS,RT,RTS,DIF,ASH(100),TJ,PCT,NBARS
      COMMON/FIVE/P(10),AS(50),XS(50),SSN(50),SSS(50),AXS(50),BARFY,STR
      COMMON/SIX/X(100),CSN(100),CSS(100),ACC(100),ATOT(100),SSH(100)
      COMMON/SEVEN/RSSH(100),XX(100),RXX(100),AM(50),XXS(100),RXXS(100)
      COMMON/EIGHT/TA(100),RTA(100),TACC(100),PACC(100),RX(100),ES,EC,IK
```

```
      COMMON/NINE/RASH(100),RATOT(100),RTOT(100),RTACC(100),TOT(100)
      AMSHELL=0.0
      AMSHT=0.0
      DO 300 I=1,K
  300 AMSHELL=AMSHELL+SSH(I)*ASH(I)*XX(I)
      IF (DNA-OD) 400,502,600
  400 DO 501 I=1,KA
  501 AMSHT=AMSHT+RSSH(I)*RASH(I)*(RX(I)+DNA-RO)
  502 AMSHELL=AMSHELL+AMSHT+G*(F/2.+DNA-RO)
  600 RETURN
      END
      SUBROUTINE COREP
      COMMON/ONE/PCON,AMCON,PSTEEL,FPC,PSHELL,AMSHELL,AC,PSHT,LL(50)
      COMMON/TWO/PCORE,AMCORE,ISHAPE,WIDTH,OD,RO,RC,DC,DT,T,TT,Z,COVER
      COMMON/THREE/NROWS,NP,FC,K,DEL,PHI,RINT,RINTS,DNA,C,F,G,TUBEFY
      COMMON/FOUR/KA,II,KK,PCAP,RCS,ROS,RT,RTS,DIF,ASH(100),IJ,PCT,NBARS
      COMMON/FIVE/P(10),AS(50),XS(50),SSN(50),SSS(50),AXS(50),BARFY,STR
      COMMON/SIX/X(100),CSN(100),CSS(100),ACC(100),ATOT(100),SSH(100)
      COMMON/SEVEN/RSSH(100),XX(100),RXX(100),AM(50),XXS(100),RXXS(100)
      COMMON/EIGHT/TA(100),RTA(100),TACC(100),RACC(100),RX(100),ES,EC,IK
      COMMON/NINE/RASH(100),RATOT(100),RTOT(100),RTACC(100),TOT(100)
      COMMON/TEN/NN,S(100),CST(100),CB(100),H(100),RCX(100)
      COMMON/ELEVEN/CX(100),SN(100),CS(100),B(100),BS(100)
      CDEL=DT/20.
      NN=INT(DNA/CDEL)
      CC=DNA-FLOAT(NN)*CDEL
      CF=CDEL-CC
      IF(DNA-Z)10,10,20
   10 CX(1)=CC+.5*CDEL
      GOTO 600
   20 CX(1)=(DNA-Z)+.5*CDEL
  600 DO 90 I=1,100
      IF(I.GT.1) CX(I)=CX(1)+(I-1)*CDEL
      SN(I)=CX(I)*PHI
      CST(I)=ES*SN(I)
      IF(CST(I).GT.TUBEFY) CST(I)=TUBEFY
      IF(CST(I).LT.-TUBEFY) CST(I)=-TUBEFY
   90 CONTINUE
      PCT=0.0
      PCORE=0.0
      IF(DNA-Z) 1,1,5
    1 IF(DNA-DIF) 80,80,4
    4 IK=INT((DNA-DIF)/CDEL)
      GOTO 6
    5 IK=20
    6 DO 70 I=1,IK
      CB(I)=CX(I)+RO-DNA
      CS(I)=CB(I)*CB(I)
      IF(RT-ABS(CB(I))) 15,15,25
   15 TOT(I)=0.0
      GOTO 30
   25 TOT(I)=2.*SQRT(RTS-CS(I))*CDEL
   30 IF (RC-ABS(CB(I))) 40,40,50
   40 TACC(I)=0.0
      GOTO 60
   50 TACC(I)=2.*SQRT(RCS-CS(I))*CDEL
   60 TA(I)=TOT(I)-TACC(I)
   70 PCORE=PCORE+TA(I)*CST(I)
      IF(DNA-Z) 80,200,200
   80 KK=INT((Z-DNA)/CDEL)
      IF(DNA-DIF) 95,100,100
   95 RCX(1)=DIF-DNA+.5*CDEL
```

```
        KK=20
        GOTO 96
100     RCX(1)=CF+.5*CDEL
        KK=INT((Z-DNA)/CDEL)
96      DO 180 I=1,KK
        IF(I.GT.1) RCX(I)=RCX(I)+(I-1)*CDEL
        B(I)=RO-DNA-RCX(I)
        BS(I)=B(I)*B(I)
        IF(RT-ABS(B(I))) 120,130,130
120     RTOT(I)=0.0
        GOTO 140
130     RTOT(I)=2.*SQRT(RTS-BS(I))*CDEL
140     IF(RC-ABS(B(I))) 150,150,160
150     RTACC(I)=0.0
        GOTO 170
160     RTACC(I)=2.*SQRT(RCS-BS(I))*CDEL
170     RTA(I)=RTOT(I)-RTACC(I)
        H(I)=CST(I)*RCX(I)/CX(I)
        IF(H(I).GT.TUBEFY) H(I)=TUBEFY
        IF(H(I).LT.-TUBEFY) H(I)=-TUBEFY
180     PCT=PCT+RTA(I)*H(I)
        PCORE=PCORE-PCT
200     RETURN
        END
        SUBROUTINE COREM
        COMMON/TWO/PCORE,AMCORE,ISHAPE,WIDTH,OD,RO,RC,DC,DT,T,TT,Z,COVER
        COMMON/THREE/NROWS,NP,FC,K,DEL,PHI,RINT,RINTS,DNA,C,F,G,TUBEFY
        COMMON/FOUR/KA,II,KK,PCAP,RCS,ROS,RT,RTS,DIF,ASH(100),TJ,PCT,NBARS
        COMMON/EIGHT/TA(100),RTA(100),TACC(100),PACC(100),RX(100),ES,EC,IK
        COMMON/NINE/RASH(100),RATOT(100),RTOT(100),RTACC(100),TOT(100)
        COMMON/TEN/NN,S(100),CST(100),CB(100),H(100),RCX(100)
        COMMON/ELEVEN/CX(100),SN(100),CS(100),B(100),BS(100)
        AMCORE=0.0
        AMCT=0.0
        IF(DNA-Z) 1,1,4
1       IF (DNA-DIF) 6,6,4
4       DO 5 I=1,IK
5       AMCORE=AMCORE+CST(I)*TA(I)*CB(I)
        IF(DNA-Z) 6,9,9
6       DO 7 I=1,KK
7       AMCT=AMCT+H(I)*RTA(I)*(RCX(I)+DNA-RO)
        AMCORE=AMCORE+AMCT
9       RETURN
        END
```

A3.4 LISTING OF INPUT FOR EXAMPLE PROBLEMS

Coding Form for PMEIX

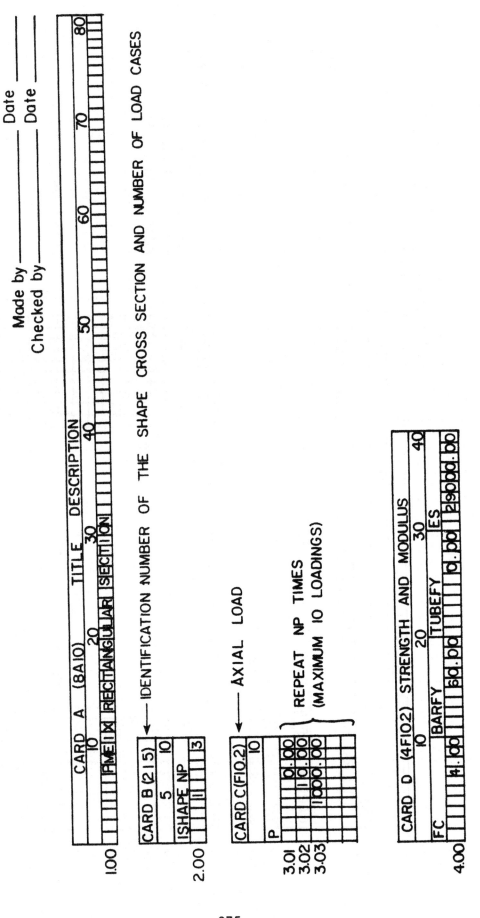

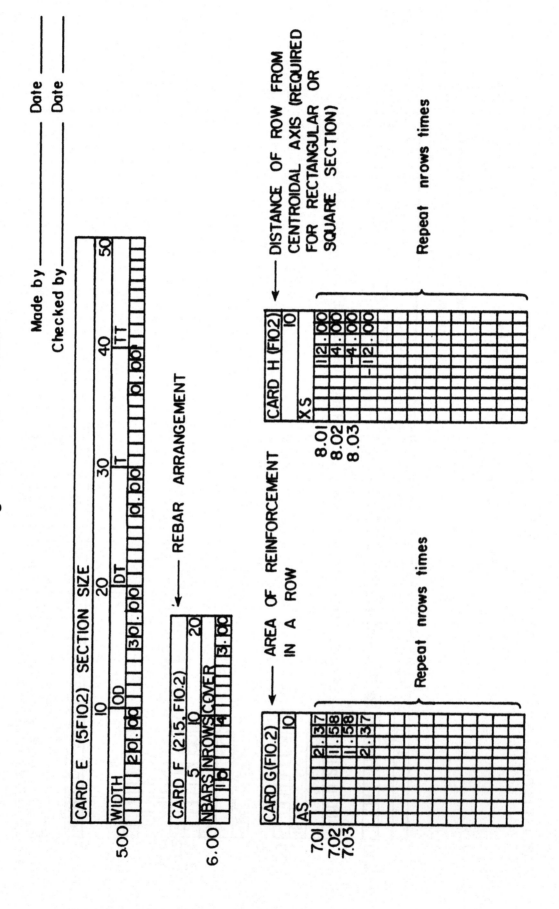

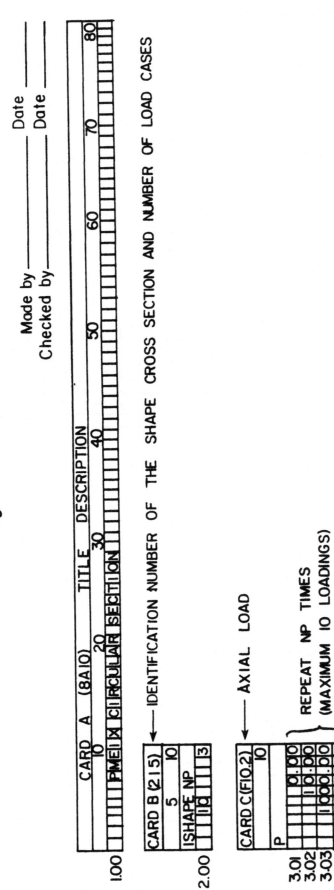

Coding Form for PMEIX

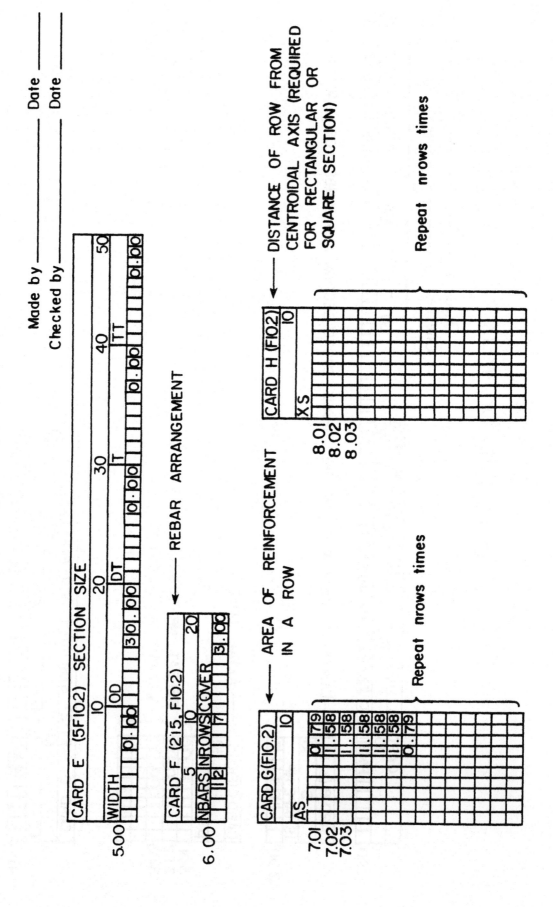

Coding Form for PMEIX

Made by _____ Date _____
Checked by _____ Date _____

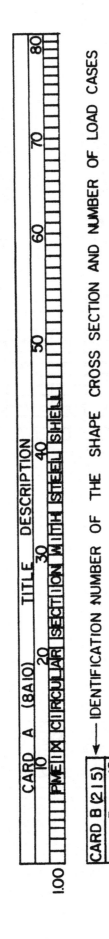
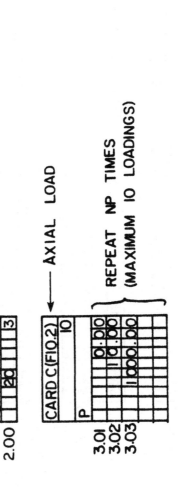

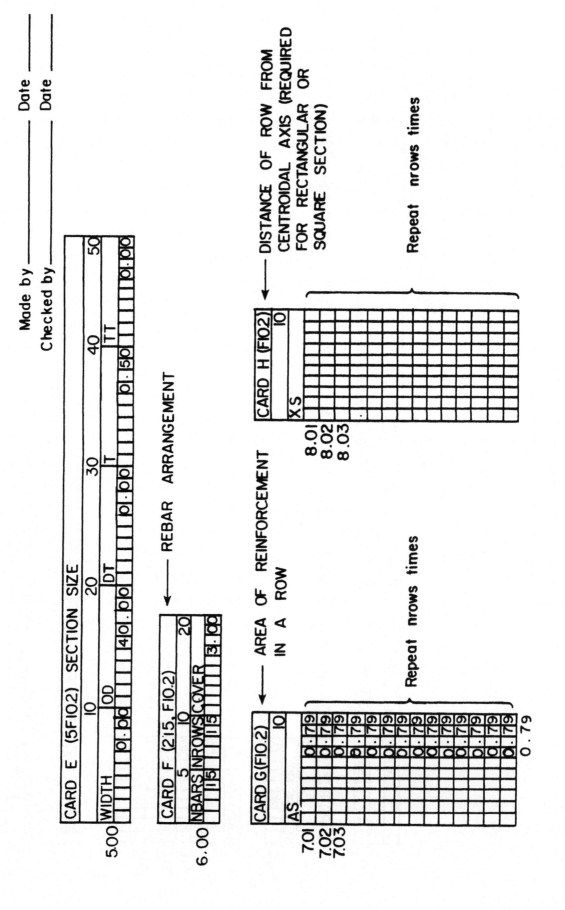

A3.5 OUTPUT OF EXAMPLE PROBLEMS

```
                                              PMEIX RECTANGULAR SECTION
        SHAPE : RECTANGULAR
        WIDTH             20.00      DEPTH                  30.00

        NO. OF REBARS                  10
        ROWS OF REBARS                  4
        COVER
        (BAR CENTER TO CONCR EDGE)    3.0

        LAYER         AREA         ORDINATE
          1           2.37           12.00

          2           1.58            4.00

          3           1.58           -4.00

          4           2.37          -12.00

        CONCRETE CYLINDER STRENGTH              4.00KSI
        REBARS YIELD STRENGTH                  60.00KSI
        SHELL/TUBE YIELD STRENGTH                  0KSI
        MODULUS OF ELAST. OF STEEL          29000.00KSI
        MODULUS OF ELAST. OF CONCR           3636.62KSI
        SQUASH LOAD CAPACITY                 2842.40KPS
```

AXIAL LOAD = 0 KIPS

MOMENT IN KIPS	EI KIP-IN2	PHI	MAX STR IN/IN	N AXIS IN
45.6	45598467.3	.000001	.00001	6.68
227.8	45591875.2	.000005	.00003	6.69
409.6	45506621.6	.000009	.00006	6.70
591.0	45460110.9	.000013	.00009	6.72
772.0	45412349.1	.000017	.00011	6.73
952.7	45364750.5	.000021	.00014	6.74
1132.9	45317276.1	.000025	.00017	6.75
1312.8	45268475.0	.000029	.00020	6.76
1492.3	45219735.6	.000033	.00022	6.78
1671.3	45171013.6	.000037	.00025	6.79
1850.0	45120804.9	.000041	.00028	6.80
2028.2	45070729.2	.000045	.00031	6.82
2205.9	45019127.3	.000049	.00033	6.83
2383.3	44967412.7	.000053	.00036	6.84
3698.7	44562741.6	.000083	.00050	6.95
4710.9	41689006.2	.000113	.00078	6.88
5106.3	35708212.8	.000143	.00094	6.54
5412.7	31287330.2	.000173	.00109	6.29
5517.0	27181908.9	.000203	.00122	5.99
5613.7	24003138.1	.000233	.00134	5.74
5706.8	21698771.7	.000263	.00146	5.55
5790.0	19791651.8	.000293	.00159	5.43
5884.0	18216761.6	.000323	.00172	5.31
5906.0	16900944.3	.000353	.00184	5.23
5985.5	15627890.3	.000383	.00195	5.09
5998.0	14523061.9	.000413	.00206	5.00
6001.2	13546701.5	.000443	.00216	4.88
6003.5	12692490.1	.000473	.00226	4.77
6005.4	11939257.9	.000503	.00235	4.68
6006.7	11269678.2	.000533	.00245	4.60
6008.2	10671838.5	.000563	.00245	4.54
5971.0	10070538.6	.000593	.00267	4.50
6014.2	9653574.6	.000623	.00279	4.48
6013.3	9208694.8	.000653	.00289	4.42
6012.5	8803068.2	.000683	.00298	4.37
6011.6	8431400.0	.000713	.00308	4.32
6010.6	8089628.9	.000743	.00318	4.28
6009.2	7773870.0	.000773	.00328	4.24
6008.0	7481909.6	.000803	.00338	4.21
6006.4	7210589.9	.000833	.00348	4.18
6005.2	6958544.4	.000863	.00358	4.15
6003.8	6723162.9	.000893	.00368	4.12
6002.4	6503128.9	.000923	.00378	4.10
6001.0	6296926.2	.000953	.00389	4.08

AXIAL LOAD =   10.00 KIPS

MOMENT IN KIPS	EI KIP-IN2	PHI	MAX STR IN/IN	N AXIS IN
119.7	119744269.2	.000001	.00002	15.26
309.9	61982252.7	.000005	.00004	8.94
491.9	54656694.4	.000009	.00007	8.01
673.2	51787693.3	.000013	.00010	7.64
854.2	50244373.2	.000017	.00013	7.46
1034.6	49264755.2	.000021	.00015	7.33
1214.6	48583935.5	.000025	.00018	7.25
1394.2	48076282.7	.000029	.00021	7.20
1573.5	47680357.4	.000033	.00024	7.16
1752.3	47358516.1	.000037	.00026	7.13
1930.6	47088745.2	.000041	.00029	7.11
2108.6	46856885.3	.000045	.00032	7.10
2286.1	46654895.3	.000049	.00035	7.09
2463.2	46476391.8	.000053	.00038	7.08
3776.7	45502850.2	.000083	.00059	7.10
4806.6	42536071.5	.000113	.00079	7.01
5202.7	36382228.0	.000143	.00095	6.66
5521.8	31917668.0	.000173	.00111	6.41
5623.2	27700474.3	.000203	.00123	6.07
5723.0	24562042.0	.000233	.00136	5.84
5815.2	22110968.3	.000263	.00148	5.65
5874.8	20047947.2	.000293	.00161	5.50
5991.7	18550067.9	.000323	.00175	5.41
6072.6	17202961.3	.000353	.00188	5.32
6102.2	15932695.2	.000383	.00199	5.18
6108.3	14789990.3	.000413	.00209	5.05
6119.1	13812774.2	.000443	.00221	4.98
6120.7	12940145.4	.000473	.00231	4.87
6122.4	12171768.5	.000503	.00240	4.78
6123.6	11488849.2	.000533	.00250	4.70
6124.2	10877717.6	.000563	.00261	4.63
6124.8	10328469.0	.000593	.00271	4.57
6125.8	9831464.8	.000623	.00281	4.51
6103.5	9346899.9	.000653	.00204	4.50
6129.8	8974886.4	.000683	.00305	4.47
6128.6	8595511.9	.000713	.00315	4.42
6127.2	8246568.5	.000743	.00325	4.38
6125.7	7924556.1	.000773	.00335	4.34
6124.3	7626725.3	.000803	.00345	4.30
6122.8	7350315.2	.000833	.00356	4.27
6121.4	7093201.9	.000863	.00366	4.24
6119.9	6853154.7	.000893	.00376	4.21
6118.4	6628831.3	.000923	.00386	4.18

AXIAL LOAD = 1000.00 KIPS

MOMENT IN KIPS	EI KIP-IN2	PHI	MAX STR IN/IN	N AXIS IN
173.8	172952431.9	.000001	.00001	411.19
864.5	172805960.6	.000005	.00047	94.35
1554.9	172764255.6	.000009	.00053	59.22
2243.2	172556647.0	.000013	.00060	45.77
2928.6	172273482.1	.000017	.00066	38.69
3610.2	171914478.3	.000021	.00072	34.35
4286.9	171477986.5	.000025	.00079	31.42
4951.9	170756800.0	.000029	.00085	29.32
5523.3	167371739.7	.000033	.00091	27.66
6011.6	162475920.9	.000037	.00097	26.29
6434.5	156940066.5	.000041	.00103	25.14
6810.5	151344211.0	.000045	.00109	24.15
7149.9	145916227.2	.000049	.00114	23.31
7455.7	140673734.0	.000053	.00120	22.56
9185.5	110668267.6	.000083	.00158	19.02
10373.1	91747543.0	.000113	.00195	17.28
11235.7	78571536.5	.000143	.00234	16.37
11807.2	68249744.6	.000173	.00276	15.97
11959.2	58912318.3	.000203	.00319	15.74
11917.2	51146734.1	.000233	.00362	15.54
11843.8	45033612.9	.000263	.00407	15.47

CIRCULAR SECTION PMEIX

```
SHAPE : CIRCULAR
DIAMETER                    30.00
SHELL THICKNESS                 0
CORE TUBE O.D.                  0
CORE TUBE THICKNESS             0

NO. OF REBARS                  12
ROWS OF REBARS                  7
COVER
(BAR CENTER TO CONCR EDGE)    3.0

    LAYER       AREA        ORDINATE
      1          .79          12.00
      2         1.58          10.39
      3         1.58           6.00
      4         1.58           0
      5         1.58          -6.00
      6         1.58         -10.39
      7          .79         -12.00

CONCRETE CYLINDER STRENGTH       4.00KSI
REBARS YIELD STRENGTH           60.00KSI
SHELL/TUBE YIELD STRENGTH           0KSI
MODULUS OF ELAST. OF STEEL   29000.00KSI
MODULUS OF ELAST. OF CONCR    3636.62KSI
SQUASH LOAD CAPACITY          3358.32KPS
```

AXIAL LOAD = 0 KIPS

MOMENT IN KIPS	EI KIP-IN2	PHI	MAX STR IN/IN	N AXIS IN
43.1	43133282.0	.000001	.00001	7.77
215.4	43082893.4	.000005	.00004	7.79
387.3	43031378.1	.000009	.00007	7.80
558.7	42980128.1	.000013	.00010	7.81
729.8	42927717.8	.000017	.00013	7.82
900.4	42875515.6	.000021	.00016	7.83
1070.6	42822118.8	.000025	.00020	7.84
1240.3	42768872.1	.000029	.00023	7.85
1409.6	42715733.8	.000033	.00026	7.87
1578.5	42661336.3	.000037	.00029	7.88
1746.8	42605673.5	.000041	.00032	7.89
1914.8	42550041.1	.000045	.00036	7.90
2082.2	42494396.3	.000049	.00039	7.92
2249.2	42437418.9	.000053	.00042	7.93
3486.3	42004147.3	.000083	.00066	8.01
4664.5	41278908.1	.000113	.00091	8.09
5259.5	36779714.6	.000143	.00113	7.91
5587.9	32300283.8	.000173	.00132	7.65
5775.0	28448323.9	.000203	.00151	7.43
5940.7	25535403.7	.000233	.00169	7.24
6115.8	23250859.9	.000263	.00187	7.11
6162.3	21031741.3	.000293	.00204	6.95
6197.7	19188064.1	.000323	.00219	6.79
6230.8	17648655.0	.000353	.00235	6.66
6259.9	16344300.9	.000383	.00251	6.56
6263.5	15165743.9	.000413	.00268	6.50
6316.3	14258024.3	.000443	.00286	6.45
6339.4	13402459.1	.000473	.00302	6.39
6361.7	12647543.2	.000503	.00319	6.34
6382.9	11975473.1	.000533	.00336	6.30
6403.2	11373383.2	.000563	.00352	6.26
6422.7	10830842.9	.000593	.00370	6.23
6441.9	10340056.5	.000623	.00387	6.21

AXIAL LOAD = 10.00 KIPS

MOMENT IN KIPS	EI KIP-IN2	PHI	MAX STR IN/IN	N AXIS IN
106.9	106885876.9	.000001	.00002	15.21
286.7	57338638.1	.000005	.00005	9.77
458.9	50987794.0	.000009	.00008	8.97
630.2	48475489.9	.000013	.00011	8.63
801.1	47124746.7	.000017	.00014	8.47
971.5	46261566.6	.000021	.00018	8.36
1141.4	45657713.1	.000025	.00021	8.28
1311.9	45206470.9	.000029	.00024	8.24
1480.1	44850073.1	.000033	.00027	8.20
1648.7	44559405.1	.000037	.00030	8.18
1816.8	44313274.1	.000041	.00033	8.16
1984.5	44100872.1	.000045	.00037	8.15
2151.8	43914480.7	.000049	.00040	8.14
2318.6	43746626.9	.000053	.00043	8.14
3553.5	42812792.8	.000083	.00068	8.16
4734.9	41901395.3	.000113	.00093	8.21
5324.8	37236626.1	.000143	.00114	8.00
5678.3	32822520.7	.000173	.00134	7.75
5843.5	28785956.1	.000203	.00152	7.50
6039.1	25918777.1	.000233	.00171	7.33
6205.8	23503343.7	.000263	.00189	7.20
6254.3	21345756.2	.000293	.00205	7.01
6295.2	19489659.6	.000323	.00222	6.88
6326.5	17922198.8	.000353	.00238	6.75
6355.5	16594094.2	.000383	.00254	6.64
6382.1	15453149.0	.000413	.00271	6.56
6371.6	14382828.5	.000443	.00288	6.50
6434.3	13603164.2	.000473	.00306	6.47
6456.8	12835067.6	.000503	.00323	6.42
6476.4	12150775.6	.000533	.00340	6.37
6496.2	11538514.7	.000563	.00357	6.34
6515.6	10987438.6	.000593	.00374	6.31
6534.2	10488263.6	.000623	.00391	6.28

AXIAL LOAD = 1000.00 KIPS

MOMENT IN KIPS	EI KIP-IN2	PHI	MAX STR IN/IN	N AXIS IN
160.3	160290079.8	.000001	.00034	344.56
801.3	160254026.5	.000005	.00040	80.99
1441.5	160169787.8	.000009	.00047	51.76
2080.5	160037389.9	.000013	.00053	40.56
2717.6	159856612.0	.000017	.00059	34.67
3352.2	159627403.5	.000021	.00065	31.04
3973.1	158923405.5	.000025	.00071	28.58
4524.9	156030244.0	.000029	.00078	26.76
5003.7	151627338.1	.000033	.00084	25.32
5421.8	146536170.3	.000037	.00089	24.14
5795.2	141345258.2	.000041	.00095	23.16
6133.4	136297141.6	.000045	.00101	22.33
6438.0	131387060.9	.000049	.00106	21.61
6725.8	126902424.4	.000053	.00111	20.99
8371.1	100856084.5	.000083	.00149	17.99
9569.0	84681519.6	.000113	.00186	16.46
10507.2	73476574.8	.000143	.00223	15.58
11281.7	65212256.0	.000173	.00261	15.07
11708.4	57676720.1	.000203	.00299	14.72
11833.2	50786119.4	.000233	.00338	14.50
11909.3	45282605.0	.000263	.00377	14.35
11960.6	40821305.8	.000293	.00418	14.26

CIRC. SECT. W/STEEL SHELL PMEIX

```
SHAPE : CIRCULAR
DIAMETER                    40.00
SHELL THICKNESS               .50
CORE TUBE O.D.                 0
CORE TUBE THICKNESS            0

NO. OF REBARS                 15
ROWS OF REBARS                15
COVER
(BAR CENTER TO CONCR EDGE)   3.0

LAYER         AREA        ORDINATE
  1            .79          16.41
  2            .79          15.69
  3            .79          14.29
  4            .79          12.26
  5            .79           9.70
  6            .79           6.71
  7            .79           3.43
  8            .79           0
  9            .79          -3.43
 10            .79          -6.71
 11            .79          -9.70
 12            .79         -12.26
 13            .79         -14.29
 14            .79         -15.69
 15            .79         -16.41

CONCRETE CYLINDER STRENGTH         4.00KSI
REBARS YIELD STRENGTH             60.00KSI
SHELL/TUBE YIELD STRENGTH         36.00KSI
MODULUS OF ELAST. OF STEEL     29000.00KSI
MODULUS OF ELAST. OF CONCR      3636.62KSI
SQUASH LOAD CAPACITY            7675.65KPS
```

AXIAL LOAD =   0 KIPS

MOMENT IN KIPS	EI KIP-IN2	PHI	MAX STR IN/IN	N AXIS IN
567.1	557051870.6	.000001	.00001	15.14
2788.4	556088074.2	.000005	.00007	15.17
4996.0	555108522.3	.000009	.00013	15.19
7203.6	554123963.2	.000013	.00019	15.22
9403.1	553123107.7	.000017	.00025	15.24
11590.4	552116316.2	.000021	.00031	15.27
13777.4	551097799.2	.000025	.00037	15.30
15951.8	550062149.4	.000029	.00043	15.32
18118.6	549048553.8	.000033	.00049	15.34
20275.0	547990747.2	.000037	.00055	15.37
22424.4	546937331.0	.000041	.00061	15.40
24563.7	545860014.8	.000045	.00067	15.42
26603.9	544774096.0	.000049	.00073	15.45
28641.9	540790740.9	.000053	.00079	15.45
36507.3	440930906.8	.000083	.00116	14.44
40305.0	357478106.4	.000113	.00149	13.67
41900.1	293700214.6	.000143	.00181	13.13
42900.1	247977607.6	.000173	.00211	12.71
43470.1	214184383.5	.000203	.00242	12.42
43840.3	188194597.5	.000233	.00272	12.17
44035.2	167396150.6	.000263	.00302	12.00
44250.9	151026917.2	.000293	.00334	11.90
44337.6	137267960.8	.000323	.00365	11.81
44304.4	125763244.9	.000353	.00396	11.73

AXIAL LOAD =   10.00 KIPS

MOMENT IN KIPS	EI KIP-IN2	PHI	MAX STR IN/IN	N AXIS IN
605.3	605252734.4	.00001	.00002	17.56
2828.5	565706129.4	.00005	.00008	15.67
5083.6	560399446.7	.00009	.00013	15.47
7258.7	557747843.1	.00013	.00019	15.41
9449.9	555874043.1	.00017	.00025	15.39
11640.7	554319493.2	.00021	.00031	15.38
13823.3	552930322.5	.00025	.00037	15.39
15907.4	551635260.0	.00029	.00043	15.40
18163.0	550393096.0	.00033	.00049	15.42
20319.7	549180295.7	.00037	.00055	15.44
22467.6	547989546.4	.00041	.00061	15.46
24606.4	546809117.4	.00045	.00067	15.48
26736.0	545632748.5	.00049	.00074	15.51
28718.1	541850616.1	.00053	.00080	15.50
36682.0	441951772.9	.00083	.00116	14.49
40484.2	358267490.0	.00113	.00149	13.72
42090.1	294330317.0	.00143	.00181	13.17
42907.2	248515769.6	.00173	.00212	12.75
43575.1	214655827.2	.00203	.00243	12.46
43984.4	188602775.3	.00233	.00273	12.21
44206.0	168083606.1	.00263	.00304	12.04
44303.3	151342293.2	.00293	.00336	11.95
44429.1	137551481.3	.00323	.00367	11.85
44483.4	126015274.7	.00353	.00398	11.78

AXIAL LOAD = 1000.00 KIPS

MOMENT IN KIPS	EI KIP-IN2	PHI	MAX STR IN/IN	N AXIS IN
862.9	862882847.9	.00001	.00016	160.83
4313.8	862766058.4	.00005	.00024	48.26
7705.3	856140012.3	.00009	.00032	35.76
10569.9	813071226.5	.00013	.00039	30.61
13083.8	769632820.2	.00017	.00046	27.65
15439.0	735192733.5	.00021	.00053	25.71
17700.4	708177710.3	.00025	.00060	24.33
19930.6	686916910.2	.00029	.00066	23.31
22002.3	669462244.2	.00033	.00073	22.51
24239.1	655110757.4	.00037	.00079	21.88
26357.7	642871721.6	.00041	.00086	21.36
28445.0	632554715.5	.00045	.00092	20.94
30558.5	623643064.7	.00049	.00098	20.59
32637.9	615620003.1	.00053	.00105	20.29
43017.5	518283343.6	.00083	.00151	18.73
47235.7	418014986.6	.00113	.00196	17.84
49260.0	344475861.8	.00143	.00239	17.21
50245.3	290435537.8	.00173	.00281	16.76
50750.7	250003288.1	.00203	.00325	16.51
50945.8	218737419.3	.00233	.00368	16.31
51060.0	194139975.9	.00263	.00413	16.20

APPENDIX 4

FORMS FOR MAKING SOLUTIONS USING NONDIMENSIONAL METHOD

A4.1 NONDIMENSIONAL ANALYSIS OF LATERALLY LOADED PILES WITH PILE HEAD RESTRAINED AGAINST ROTATION

$P_t = \underline{}$ lb $\qquad EI = \underline{}$ lb-in^2

Trial $k_{assumed} = \underline{}$ lb-in^3 (or $T_{assumed} = \underline{}$ in)

$T = \left(\dfrac{EI}{k}\right)^{1/5} = \underline{}$ in $\qquad k_\theta = \underline{} \dfrac{\text{in-lb}}{\text{radian}} \qquad A_{st} = \underline{} \qquad B_{st} = \underline{}$

$$M_t = \dfrac{k_\theta A_{st} P_t T^2}{EI} \bigg/ \left(1 - \dfrac{B_{st} k_\theta T}{EI}\right) = \underline{} \text{ in-lb}$$

Depth	Depth Coefficient	Deflection Coefficient	Deflection Coefficient	Deflection	Soil Resistance	Soil Modulus
in				in	lb/in	lb/in^2
x	z $= \dfrac{x}{T}$	A_y from Fig.	B_y from Fig.	y $= A_y \dfrac{P_t T^3}{EI} + B_y \dfrac{M_t T^2}{EI}$	p from p-y curve	E_s $= -\dfrac{p}{y}$

$k = \dfrac{E_s}{x} = \underline{} \qquad\qquad T_{obtained} = \left(\dfrac{EI}{k}\right)^{1/5} = \underline{}$ in

A4.2 NONDIMENSIONAL ANALYSIS OF LATERALLY LOADED PILES WITH PILE HEAD FIXED AGAINST ROTATION

$P_t =$ _____ lb $M_t =$ _____ in-lb $EI =$ _____ lb-in^2

Trial _____ $k_{assumed} =$ _____ lb/in^3 (or $T_{assumed} =$ _____ in)

$T = \left(\dfrac{EI}{k}\right)^{1/5} =$ _____ in $z_{max} = \dfrac{L}{T} =$ _____

Depth	Depth Coefficient	Deflection Coefficient	Deflection	Soil Resistance	Soil Modulus
in			in	lb/in	lb/in^2
x	z $= \dfrac{x}{T}$	F_y from Fig.	y $= F_y \dfrac{P_t T^3}{EI}$	P from p-y curve	E_s $= \dfrac{p}{y}$

$k = \dfrac{E_s}{x} =$ _____ lb/in^3 $T_{obtained} = \left(\dfrac{EI}{k}\right)^{1/5} =$ _____ in

A4.3 NONDIMENSIONAL ANALYSIS OF LATERALLY LOADED PILES WITH PILE HEAD FREE TO ROTATE

$P_t = $ _____ lbs $M_t = $ _____ in-lbs $EI = $ _____ lb-in^2

Trial $k_{assumed} = $ _____ lb/in^3 (or $T_{assumed} = $ _____ in)

$T = \left(\dfrac{EI}{k}\right)^{1/5} = $ _____ in $z_{max} = \dfrac{L}{T} = $ _____

Depth in x	Depth Coefficient $z = \dfrac{x}{T}$	Deflection Coefficient A_y from Fig.	Deflection Coefficient B_y from Fig.	Deflection in y $= A_y \dfrac{P_t T^3}{EI} + B_y \dfrac{M_t T^2}{EI}$	Soil Resistance lb/in p from p-y curve	Soil Modulus lb/in^2 $E_s = -\dfrac{p}{y}$

$k = \dfrac{E_s}{x} = $ _____ lb/in^3 $T_{obtained} = \left(\dfrac{EI}{k}\right)^{1/5} = $ _____ in

APPENDIX 5

ORDER FORM FOR COMPUTER PROGRAMS COM624 AND PMEIX

ORDER FORM FOR COMPUTER PROGRAMS COM624 AND PMEIX

No.	Computer Program	Author(s)	Price	Indicate Source Deck or Magnetic Tape
GS80-1	COM624	Reese and Sullivan	$350.00	_____
	PMEIX	Reese	150.00	_____

TAPE SPECIFICATION: IBM___ CDC___ No. of tracks:___ Density:___BPI
Record length:____characters/record Blocking factor:____records/block
Physical record size:_____   Recorded in: ASCII_____ EBCDIC_____
Labeled_____ Unlabeled_____   Blocked_____ Unblocked_____

    Above price includes card deck or magnetic tape, as preferred, and complete documentation with compiled listing and executed test problems, but <u>does not</u> include shipping costs. Please indicate shipping method: United States - First Class____, Fourth Class, Book Rate____, UPS____, Federal Express____; Outside United States - First Class Airmail____, Surface Mail____. Five percent sales tax will be added to orders in State of Texas.

    Some potential users may wish to purchase the documentation material at a cost of $25 prior to purchasing the deck.

Company_____
Name of Contact Person_____
Address_____
City and State_____Zip Code_____
Telephone_____
Special Instructions_____
_____
_____

Return Order Form To
Geotechnical Engineering Center
College of Engineering
The University of Texas at Austin
Cockrell Hall 6.2
Austin, Texas   78712

APPENDIX 6

MODIFICATION OF p-y CURVES FOR BATTERED PILES

APPENDIX 6

MODIFICATION OF p-y CURVES FOR BATTERED PILES

Kubo (1965) and Awoshika and Reese (1971) investigated the effect of batter on the behavior of laterally loaded piles. Kubo used model tests in sands and full-scale field experiments to obtain his results. Awoshika and Reese tested 2-in diameter piles in sand. The value of the constant showing the increase or decrease in soil resistance as a function of the angle of batter may be obtained from the line in Fig. A6.1. The "ratio of soil resistance" was obtained by comparing the groundline deflection for a battered pile with that of a vertical pile and is, of course, based purely on experiment.

The correction for batter is made as follows: (1) enter Fig. A6.1 with the angle of batter, positive or negative, and obtain a value of the ratio; (2) compute groundline deflection as if the pile were vertical; (3) multiply the deflection found in 2 by the ratio found in 1; (4) vary the strength of the soil until the deflection found in 3 is obtained; and (5) use the modified strength found in 4 for the further computations of the behavior of the pile that is placed on a batter. The method outlined is obviously approximate and should be used with caution. If the project is large, it could be desirable to perform a field test on a pile installed with a batter.

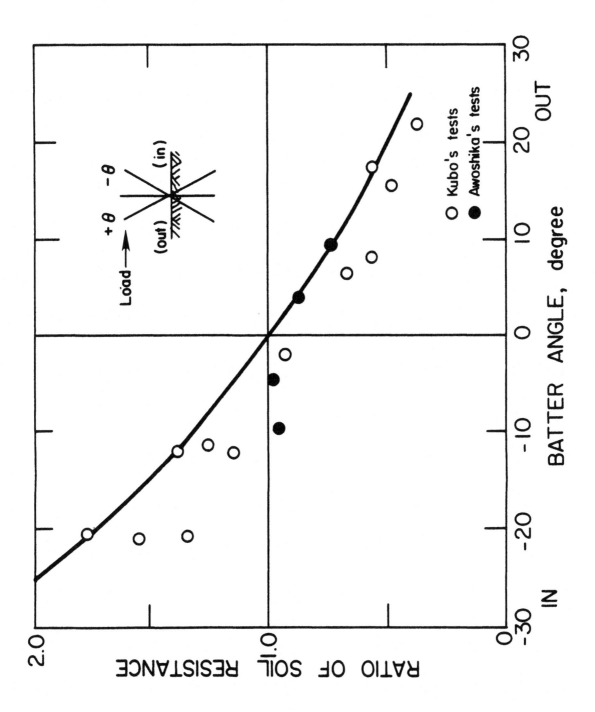

Fig. A6.1. Modification of p-y curves for battered piles (after Kubo, and Awoshika and Reese)

APPENDIX 7

EXAMPLE PROBLEMS FOR WORKSHOPS

A7.1 EXERCISE RELATING TO MATERIAL PRESENTED IN CHAPTER 2

CONSTRUCTION OF p-y CURVES

<u>Problem Statement:</u> Given the soil and pile information below, construct the appropriate p-y relationship at a depth of 12 ft below the ground surface.

Loading

Static

Pile Characteristics

diameter = 48 in

Stiff Clay
(above water table)

c = 2100 lb/sq ft
ε_{50} = 0.005 in/in
γ = 112 lb/cu ft

Step-by-Step Solution

1. Obtain the best possible estimate of soil parameters:

    At depth 12 ft

    c = 1.05 T/sq ft = 2100 lb/sq ft

    γ = 112 lb/cu ft

    ε_{50} = 0.005

    c_{avg} = 2100 lb/sq ft

2. Compute the ultimate soil resistance per unit length of pile:

    From Eq. (1), $\quad p_u = [3 + \dfrac{\gamma'}{c_{avg}} x + \dfrac{J}{b} x]cb$

    $\quad\quad = [3 + \dfrac{112}{2100}(12) + \dfrac{0.5}{4}(12)](2100)(4)$

    $\quad\quad$ = 43,176 lb/ft

    $\quad\quad$ = 3598 lb/in

    From Eq. (2), $\quad p_u = 9cb$

    $\quad\quad$ = (9)(2100)(4)

    $\quad\quad$ = 75,600 lb/ft

    $\quad\quad$ = 6300 lb/in

    Select p_u = 3598 lb/in

3. Compute y_{50}:

    $y_{50} = 2.5\, \varepsilon_{50}\, b = (2.5)(0.005)(48) = 0.6$ in

4. Compute points describing the p-y curve:

    From Eq. (20), $\quad p = 0.5 \left(\dfrac{y}{y_{50}}\right)^{\frac{1}{4}} p_u \quad\quad 0 < y < 16y_{50}$

    $\quad\quad\quad\quad\quad\quad\; p = p_u \quad\quad\quad\quad\quad\quad\;\; y \geq 16y_{50}$

	y (in)	p (lb/in)
	0	0
	0.01	646
	0.05	967
	0.10	1149
	0.20	1367
	0.50	1719
	1.00	2044
	2.00	2431
$16y_{50}$ =>	9.60	3598

305

A7.2 EXERCISE RELATING TO MATERIAL PRESENTED IN CHAPTER 3

COMPUTER METHOD OF ANALYSIS

<u>Problem Statement, Case 1</u>: Using the information presented below, create an input data deck for the computer program COM624. In addition, have the computer program generate and print p-y values at depths of 24, 48, and 96 inches.

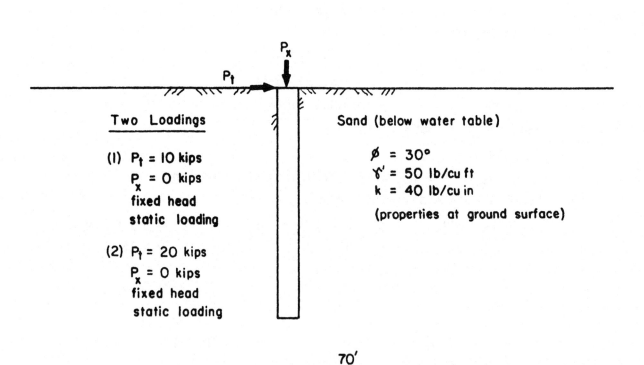

Coding Form for COM624

PROBLEM 1 Made by: _____ Date: _____
  Checked by: _____ Date: _____

1.00 TITLE CARD (18A4): `CASE STUDY STEEL PIPE PILE` (cols 1–72)

2.00 UNITS CARD (4A4): `ENGL`

3.00 INPUT CONTROL CARD (4I5): NI=1, NL=20, NDIAMNW=1, 0

4.00 INPUT CTRL. CARD (3I5): NGI=2, NSTR=2, NPY=0

5.00 PILE GEOMETRY CARD (3E10.3): LENGTH=7.20E2, EPILE=2.9OE7, XGS=0.0OE0

307

Coding Form for COM624

Made by: _____ Date: _____
Checked by: _____ Date: _____

PROBLEM 1

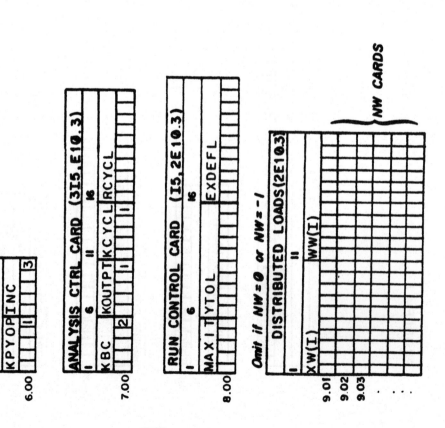

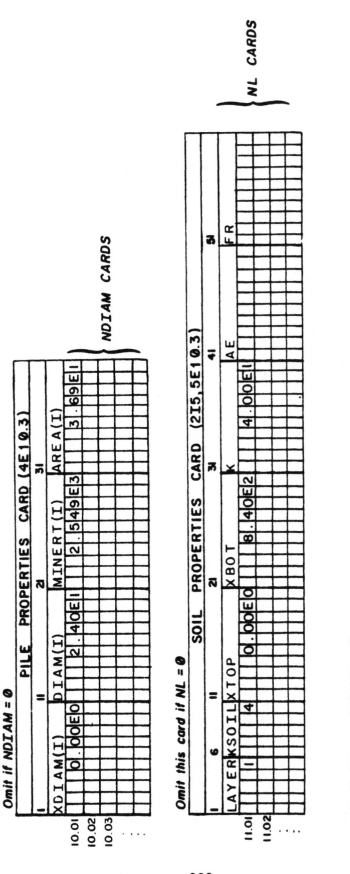

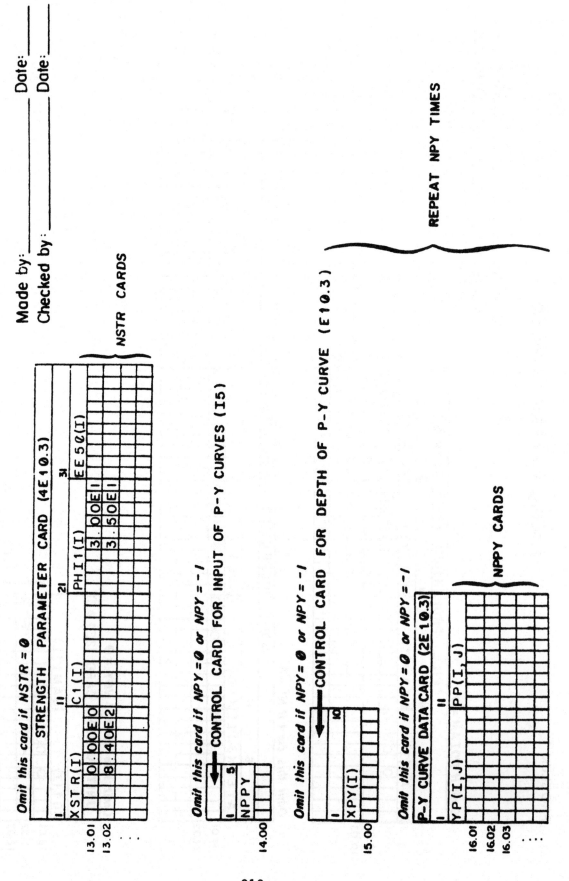

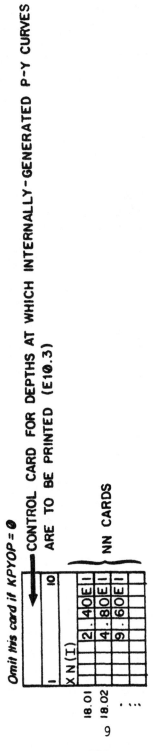

A7.3 EXERCISE RELATING TO MATERIAL PRESENTED IN CHAPTER 3

COMPUTER METHOD OF ANALYSIS

Problem Statement, Case 2: Using the information presented below, create an input data deck for the computer program COM624. In addition, have the computer program generate and print p-y values at depths of 24, 48, and 96 inches.

Loading

P_t = 10 kips
P_x = 20 kips
M = 1.0 × 10^5 in-lb
free head
cyclic loading

Sand (below water table)

ϕ = 30°
γ' = 50 lb/cu ft
k = 40 lb/cu in

(properties at ground surface)

Pile Characteristics

diameter = 24 in
length = 60 ft = 720 in
E = 3.60 × 10^6 lb/sq in
I = 16,290 in^4
Area = 452 sq in

(drilled shaft with 14 No. 8 bars on a 18-in diameter circle, f'_c = 4000 lb/sq in)

Sand (below water table)

ϕ = 35°
γ = 60 lb/cu ft
k = 40 lb/cu in

(properties at 70 ft - assume linear variation between 0 and 70 ft)

Coding Form for COM624

PROBLEM 2

Made by: _____ Date: _____
Checked by: _____ Date: _____

1.00 — TITLE CARD (18A4): `CASE STUDY DRILLED SHAFT`

2.00 — UNITS CARD (4A4): `ENGL`

3.00 — INPUT CONTROL CARD (4I5):
- NI = (blank)
- NL = 1
- NDIAM = 20
- NW = 0

4.00 — INPUT CTRL. CARD (3I5):
- NGI = 2
- NSTR = 2
- NPY = 0

5.00 — PILE GEOMETRY CARD (3E10.3):
- LENGTH = 7.20E2
- EPILE = 3.6 0E6
- XGS = 0.00E0

Coding Form for COM624

Made by: _____ Date: _____
Checked by: _____ Date: _____

PROBLEM 2

→ OUTPUT CONTROL CARD (2I5)

```
  1      6
| KPYOP | INC |
|   1   |  3  |
```
6.00

ANALYSIS CTRL CARD (3I5, E10.3)

```
  1      6      11
| KBC | KOUTPT | KCYCL | RCYCL |
|  1  |   1    |   1   |   0   |
```
7.00

RUN CONTROL CARD (I5, 2E10.3)

```
  1      6       16
| MAXIT | YTOL | EXDEFL |
```
8.00

Omit if NW=0 or NW=-1

DISTRIBUTED LOADS (2E10.3)

```
  1         11
| XW(I) | WW(I) |
```

9.01
9.02
9.03 . . . .

} NW CARDS

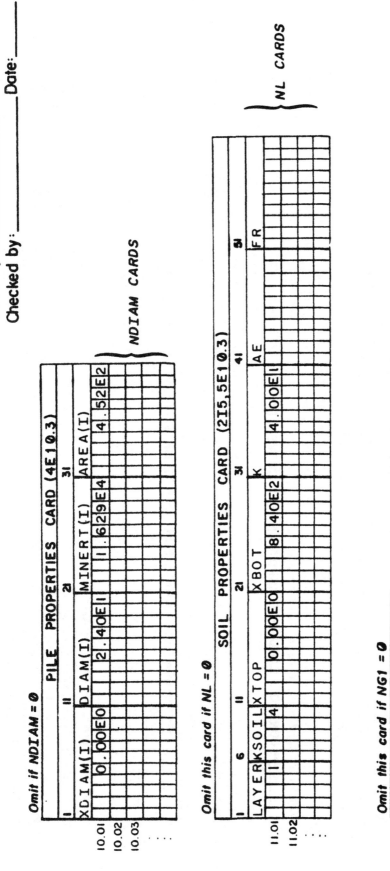

PROBLEM 2

Coding Form for COM624

Made by: _____ Date: _____
Checked by: _____ Date: _____

Omit this card if NSTR = 0

STRENGTH PARAMETER CARD (4E10.3)

	1	11	21	31
	XSTR(I)	C1(I)	PHI1(I)	EE50(I)
13.01	0.00E0		3.00E1	
13.02	8.40E2		3.50E1	
...				

} NSTR CARDS

Omit this card if NPY = 0 or NPY = -1
← CONTROL CARD FOR INPUT OF P-Y CURVES (I5)

	1 — 5
	NPPY
14.00	

Omit this card if NPY = 0 or NPY = -1
← CONTROL CARD FOR DEPTH OF P-Y CURVE (E10.3)

	1 — 10
	XPY(I)
15.00	

Omit this card if NPY = 0 or NPY = -1

P-Y CURVE DATA CARD (2E10.3)

	1	11
	YP(I,J)	PP(I,J)
16.01		
16.02		
16.03		
....		

} NPPY CARDS

} REPEAT NPY TIMES

316

Coding Form for COM624

PROBLEM 2

Made by: _____ Date: _____
Checked by: _____ Date: _____

Omit this card if KPYOP = 0
→ CONTROL CARD FOR OUTPUT OF INTERNALLY-GENERATED P-Y CURVES (I5)

17.00

1	5
I	NN
	3

Omit this card if KPYOP = 0
→ CONTROL CARD FOR DEPTHS AT WHICH INTERNALLY-GENERATED P-Y CURVES ARE TO BE PRINTED (E10.3)

1	10
I	XN(I)

18.01 2.40E1
18.02 4.80E1
...   9.60E1

} NN CARDS

CARD TO ESTABLISH LOADS ON PILE HEAD (I5, 3E10.3)

1	6	16	26
KOP	PT	BC2	PX
19.01	1	1.00E4	1.00E5	2.00E4
19.02	-1			

} MAXIMUM 20 LOADINGS

TITLE CARD (18A4)

20.00 END

A7.4 EXERCISE RELATING TO MATERIAL PRESENTED IN CHAPTER 3

COMPUTER METHOD OF ANALYSIS

Problem Statement, Case 3: Using the information presented below, create an input data deck for the computer program COM624. In addition, have the computer program generate and print p-y values at a depth of 100 inches.

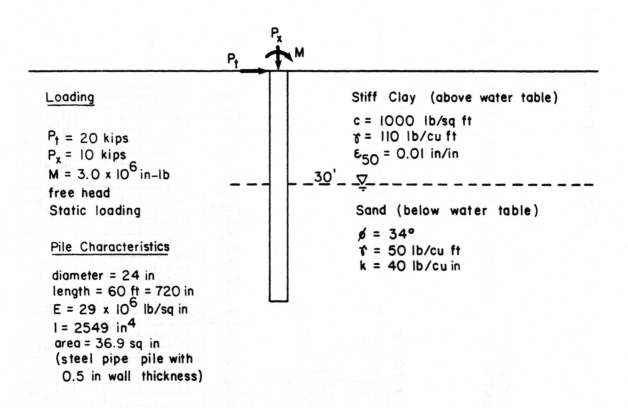

Loading

P_t = 20 kips
P_x = 10 kips
M = 3.0 x 10^6 in-lb
free head
Static loading

Pile Characteristics

diameter = 24 in
length = 60 ft = 720 in
E = 29 x 10^6 lb/sq in
I = 2549 in^4
area = 36.9 sq in
(steel pipe pile with 0.5 in wall thickness)

Stiff Clay (above water table)

c = 1000 lb/sq ft
γ = 110 lb/cu ft
ε_{50} = 0.01 in/in

Sand (below water table)

ϕ = 34°
γ = 50 lb/cu ft
k = 40 lb/cu in

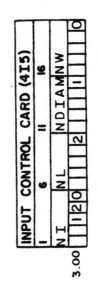

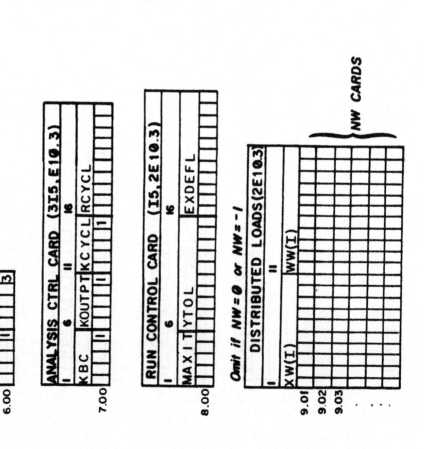

PROBLEM 3

Coding Form for COM624

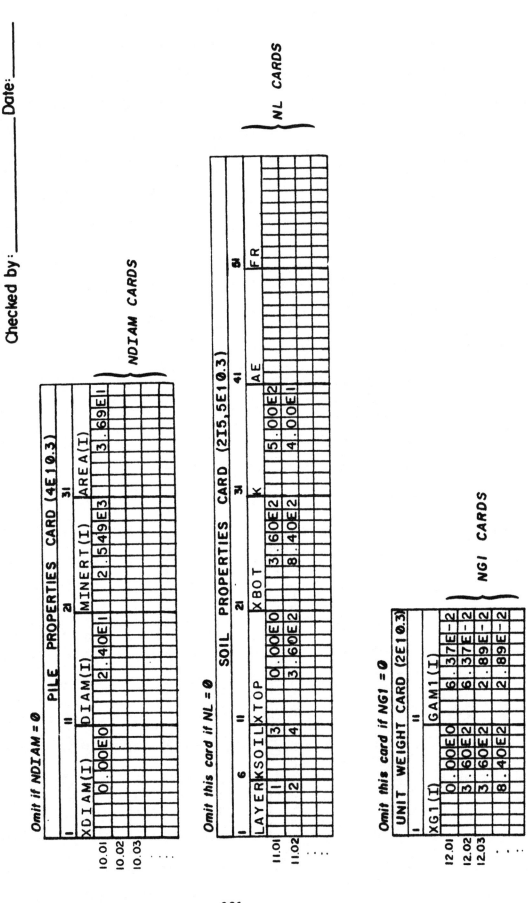

PROBLEM 3

Coding Form for COM624

Made by: _____ Date: _____
Checked by: _____ Date: _____

Omit this card if NSTR = 0

STRENGTH PARAMETER CARD (4E10.3)

	1	11	21	31	
	XSTR(I)	C1(I)	PHI1(I)	EE50(I)	
13.01	0.00E0	3.60E2	6.90E0	1.00E-2	} NSTR CARDS
13.02	3.60E2	8.40E2	6.90E0	1.00E-2	
...		3.60E2	3.40E1		
		8.40E2	3.40E1		

Omit this card if NPY = 0 or NPY = -1

→ CONTROL CARD FOR INPUT OF P-Y CURVES (I5)

	5
	NPPY
14.00	

Omit this card if NPY = 0 or NPY = -1

→ CONTROL CARD FOR DEPTH OF P-Y CURVE (E10.3)

	10
	XPY(I)
15.00	

Omit this card if NPY = 0 or NPY = -1

P-Y CURVE DATA CARD (2E10.3)

	1	11	
	YP(I,J)	PP(I,J)	
16.01			} NPPY CARDS
16.02			
16.03			
...			

} REPEAT NPY TIMES

322

Coding Form for COM624

PROBLEM 3

Made by: _____ Date: _____
Checked by: _____ Date: _____

Omit this card if KPYOP = 0
→ CONTROL CARD FOR OUTPUT OF INTERNALLY-GENERATED P-Y CURVES (I5)

17.00 | NN=1 (col 5) |

Omit this card if KPYOP = 0
→ CONTROL CARD FOR DEPTHS AT WHICH INTERNALLY-GENERATED P-Y CURVES ARE TO BE PRINTED (E10.3)

18.01 XN(I) = 1.00E2
18.02 ...
{ NN CARDS }

CARD TO ESTABLISH LOADS ON PILE HEAD (I5, 3E10.3)

KOP	PT		BC2		PX	
1		2.00E4		3.00E6		1.00E4
-1						

19.01
19.02 ....
{ MAXIMUM 20 LOADINGS }

TITLE CARD (18A4)

20.00 | END |

A7.5 EXERCISE RELATING TO MATERIAL PRESENTED IN CHAPTER 4

NONDIMENSIONAL ANALYSIS

<u>Problem Statement</u>: Using the nondimensional procedures outlined in Chapter 4, obtain the relationship between pile deflection and depth. Necessary data are given below, and the appropriate p-y curves are presented in tabular form in Table 18 (page 112).

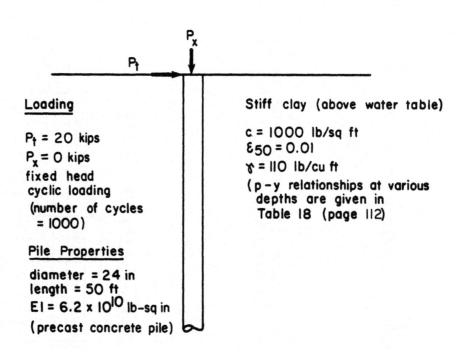

Loading

P_t = 20 kips
P_x = 0 kips
fixed head
cyclic loading
(number of cycles = 1000)

Pile Properties

diameter = 24 in
length = 50 ft
$EI = 6.2 \times 10^{10}$ lb-sq in
(precast concrete pile)

Stiff clay (above water table)

c = 1000 lb/sq ft
ε_{50} = 0.01
γ = 110 lb/cu ft
(p-y relationships at various depths are given in Table 18 (page 112)

NONDIMENSIONAL ANALYSIS OF LATERALLY LOADED PILES
WITH PILE HEAD FIXED AGAINST ROTATION

P_t = __20,000__ lb M_t = __0__ in-lb EI = __6.2 x 10^{10}__ lb-in^2

Trial __1__ $k_{assumed}$ = __18.9__ lb/in^3 (or $T_{assumed}$ = __80__ in)

$T = \left(\dfrac{EI}{k}\right)^{1/5}$ = _____ in $z_{max} = \dfrac{L}{T}$ = __600/80 = 7.5__

Depth	Depth Coefficient	Deflection Coefficient	Deflection	Soil Resistance	Soil Modulus
in			in	lb/in	lb/in^2
x	z $= \dfrac{x}{T}$	F_y from Fig. 56	y $= F_y \dfrac{P_t T^3}{EI}$	P from p-y curve	E_s $= \dfrac{p}{y}$
0	0	0.93	0.154	118	766
24	0.3	0.89	0.147	144	979
48	0.6	0.80	0.132	168	1276
96	1.2	0.53	0.088	210	2385
144	1.8	0.275	0.045	244	5422
192	2.4	0.10	0.017	201	11,852
288	3.6	-0.02	-0.003	153	46,364

$k = \dfrac{E_s}{x}$ = __29.1__ lb/in^3 $T_{obtained} = \left(\dfrac{EI}{k}\right)^{1/5}$ = __73.4__ in

NONDIMENSIONAL ANALYSIS OF LATERALLY LOADED PILES
WITH PILE HEAD FIXED AGAINST ROTATION

P_t = __20,000__ lb M_t = __0__ in-lb EI = __6.2×10^{10}__ lb-in^2

Trial __2__ $k_{assumed}$ = __0.817__ lb/in^3 (or $T_{assumed}$ = __150__ in)

$T = \left(\dfrac{EI}{k}\right)^{1/5}$ = _____ in $z_{max} = \dfrac{L}{T}$ = __600/150 = 4__

Depth	Depth Coefficient	Deflection Coefficient	Deflection	Soil Resistance	Soil Modulus
in			in	lb/in	lb/in^2
x	z $= \dfrac{x}{T}$	F_y from Fig. 56	y $= F_y \dfrac{P_t T^3}{EI}$	p from p-y curve	E_s $= \dfrac{p}{y}$
0	0	0.95	1.035	218	211
24	0.16	0.94	1.024	269	263
48	0.32	0.91	0.991	318	321
96	0.64	0.80	0.871	410	471
144	0.96	0.66	0.713	490	688
192	1.28	0.51	0.555	527	950
288	1.92	0.25	0.272	408	1501

$k = \dfrac{E_s}{x}$ = __4.17__ lb/in^3 $T_{obtained} = \left(\dfrac{EI}{k}\right)^{1/5}$ = __108__ in

NONDIMENSIONAL ANALYSIS OF LATERALLY LOADED PILES
WITH PILE HEAD FIXED AGAINST ROTATION

P_t = __20,000__ lb M_t = __0__ in-lb EI = __6.2×10^{10}__ lb-in^2

Trial __3__ $k_{assumed}$ = __65__ lb/in^3 (or $T_{assumed}$ = __62.5__ in)

$T = \left(\dfrac{EI}{k}\right)^{1/5}$ = _____ in $z_{max} = \dfrac{L}{T}$ = __600/62.5 = 9.6__

Depth	Depth Coefficient	Deflection Coefficient	Deflection	Soil Resistance	Soil Modulus
in			in	lb/in	lb/in^2
x	z $= \dfrac{x}{T}$	F_y from Fig. 56	y $= F_y \dfrac{P_t T^3}{EI}$	P from p-y curve	E_s $= \dfrac{p}{y}$
0	0	0.930	0.073		
24	0.384	0.875	0.069		
48	0.768	0.710	0.056		
96	1.536	0.380	0.030		
144	2.304	0.120	0.009		
192	3.072	0	0		
250	4.000	-0.02	-0.002		

$k = \dfrac{E_s}{x}$ = _____ lb/in^3 $T_{obtained} = \left(\dfrac{EI}{k}\right)^{1/5}$ = _____ in

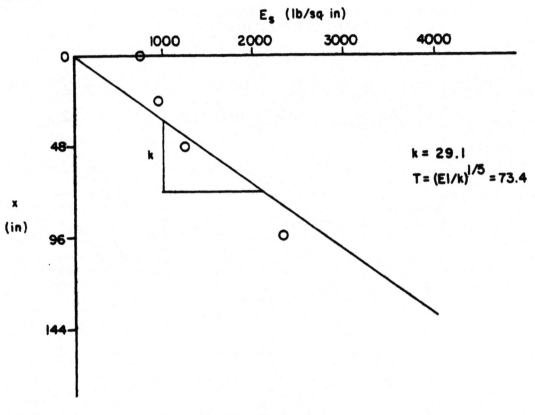
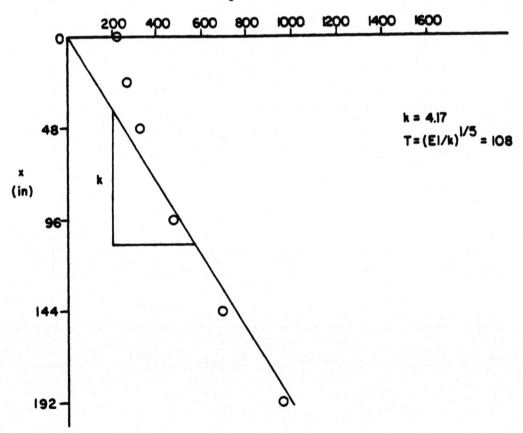

Graphic Solution for T_{final}

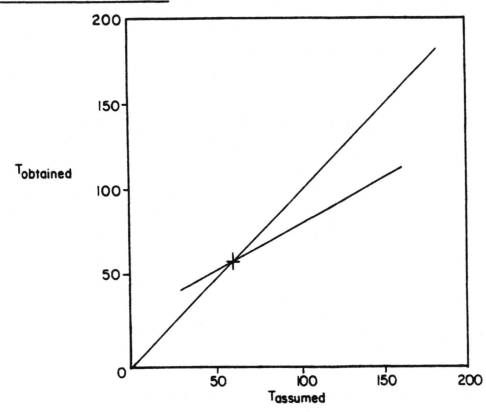

Deflection vs. depth

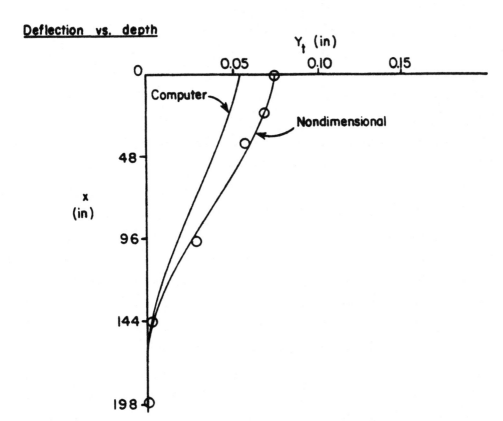

A7.6 EXERCISE RELATING TO MATERIAL PRESENTED IN CHAPTER 5

BROMS METHOD

Problem Statement: Using Broms method of analysis, solve for the maximum lateral load that may be applied to the shaft, and compute the lateral deflection at a service load of 49 kips. Information regarding significant soil and pile properties are given below.

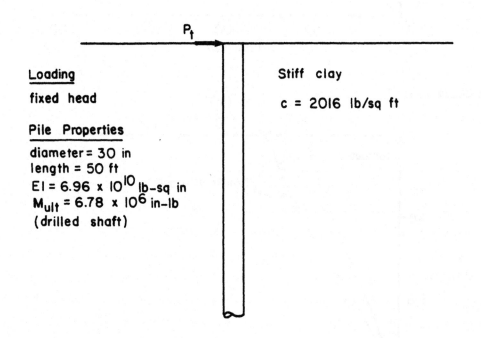

Loading
fixed head

Pile Properties
diameter = 30 in
length = 50 ft
$EI = 6.96 \times 10^{10}$ lb-sq in
$M_{ult} = 6.78 \times 10^6$ in-lb
(drilled shaft)

Stiff clay
$c = 2016$ lb/sq ft

Solution
1. The drilled shaft with 30-inch OD has 12 No. 8 rebars placed on a 24-inch diameter. The total length is 50 ft and the ultimate concrete strength (f'_c) is 4000 lb/sq in.
2. The EI is 6.96×10^{10} lb-sq in and its yield moment can be computed from computer program PMEIX which gives the value $M_{ult} = 6.78 \times 10^6$ in-lb.
3. It is assumed that the pile head is fixed against restraint.
4. The ultimate lateral load acted on the pile head at ground surface will be computed.
5. The average shear strength c is 2016 lb/sq ft.
6. The pile is not free-head.
7. Solving for the point where the pile goes from a short pile to one of intermediate length, using Eqs. 74 and 80:

From Eq. (74), $P_{ult} = 9cb(L - 1.5b)$
$= (9)(2.016)(2.5)[L - (1.5)(2.5)]$
$= 45.36L - 170.1$

$M_y = 6.78 \times 10^6$ lb-in $= 5.65 \times 10^2$ ft-kip

From Eq. (80), $P_{ult} = M_y/(0.5L + 0.75b)$
$= 5.65 \times 10^2/[0.5L + (0.75)(2.5)]$
$P_{ult}L + 3.75 P_{ult} - 1.13 \times 10^3 = 0$

Solving simultaneously
$P_{ult} = 113$ kips, $L = 6.24$ ft

Solving for the point where the pile goes from one of intermediate length to a long pile, using Eqs. 75 through 78:

From Eq. (75), $P_{ult} = \dfrac{M^{pos}_{max} + M_y}{(1.5b + 0.5f)}$

$= \dfrac{2M_y}{(1.5b + 0.5f)} = \dfrac{(2)(565)}{[(1.5)(2.5) + 0.5f]}$

From Eq. (76), $g = \left[\dfrac{M^{pos}_{max}}{2.25cb}\right]^{\frac{1}{2}} = \left[\dfrac{5.65 \times 10^2}{(2.25)(2.016)(2.5)}\right]^{\frac{1}{2}}$

$= 7.06$ ft

From Eq. (77), $L = 1.5b + f + g = (1.5)(2.5) + f + 7.06$
$= 10.81 + f$

From Eq. (78), $f = P_{ult}/9cb = P_{ult}/[(9)(2.016)(2.5)]$
$P_{ult} = 45.36f$

Solving
$$\begin{cases} P_{ult} = \dfrac{(2)(565)}{(1.5)(2.5) + 0.5f} \\ P_{ult} = 45.36f \end{cases}$$

$f = 4.24$ ft, $P_{ult} = 192.5$ kips
and $L = 10.81 + f = 10.81 + 4.24 = 15.1$ ft

These computations indicate that the pile with 50 ft pile length is in the range of long piles and the P_{ult} computed from Eq. 79 is:

$$P_{ult} = \frac{M_y}{1.5b + 0.5f} = \frac{(2)(565)}{(1.5)(2.5) + (0.5)(4.24)} = 192.5 \text{ kips}$$

The computer result of P_{ult} from Fig. 33 is 122 kips.

Deflection is computed at service load (49 kips) as following:

$\alpha = 78$ lb/sq in from Table 24

$$\beta = \left(\frac{\alpha}{4EI}\right)^{0.25} = \left(\frac{78}{(4)(6.96)(10^{10})}\right)^{0.25} = 0.0041$$

From Eq. (91), $y_t = \dfrac{P t \beta}{\alpha} A_1$

$$= \frac{(49)(10^3)(0.0041)}{78}(1) = 2.57 \text{ in.}$$

The ground deflection from the computer solution was 0.2 inches.

APPENDIX 8

ASTM STANDARDS

Reprinted, with permission from ASTM, 1916 Race Street, Philadelphia, PA 19103.

1983
ANNUAL BOOK OF ASTM STANDARDS

SECTION 4 Construction

VOLUME 04.08 Soil and Rock; Building Stones

Includes standards of the following committees:

C-18 on Natural Building Stones
D-4 on Road and Paving Materials (partial)
D-18 on Soil and Rock

Publication Code Number (PCN) 01-040883-38

 1916 Race Street/Philadelphia, PA 19103

pp. 627-646

Designation: D 3966 – 81

AMERICAN SOCIETY FOR TESTING AND MATERIALS
1916 Race St., Philadelphia, Pa. 19103
Reprinted from the Annual Book of ASTM Standards, Copyright ASTM
If not listed in the current combined index, will appear in the next edition.

Standard Method of Testing
PILES UNDER LATERAL LOADS[1]

This standard is issued under the fixed designation D 3966; the number immediately following the designation indicates the year of original adoption or, in the case of revision, the year of last revision. A number in parentheses indicates the year of last reapproval.

1. Scope

1.1 This method covers procedures for testing vertical and batter piles either individually or in groups to determine the load-deflection relationship when subjected to lateral loading. It is applicable to all deep foundation units regardless of their size or method of installation. This method is divided into the following sections:

	Section
Applicable Documents	2
Significance and Use	3
Apparatus for Applying Loads	4
Apparatus for Measuring Movements	5
Loading Procedures	6
Procedures for Measuring Movements	7
Safety Requirements	8
Report	9
Precision and Accuracy	10

1.2 This method only describes procedures for testing single piles or pile groups. It does not cover the interpretation or analysis of the test results or the application of the test results to foundation design or the use of empirical or analytic procedures for determining the magnitude and variation of the coefficient of horizontal subgrade reaction, bending stresses, and bending moments over the length of the pile. The term "failure" as used in this method indicates a rapid progressive lateral movement of the pile or pile group under a constant or decreasing load.

1.3 Apparatus and procedures designated "optional" are to be required only when included in the project specifications and, if not specified, may be used only with the approval of the engineer responsible for the foundation design. The word "shall" indicates a mandatory provision and "should" indicates a recommended or advisory provision. Imperative sentences indicate mandatory provisions. Notes and illustrations included herein are explanatory or advisory.

1.4 The values stated in inch-pound units are to be regarded as the standard.

2. Applicable Documents

2.1 *ASTM Standards*:
A 36 Specification for Structural Steel[2]
A 240 Specification for Heat-Resisting Chromium and Chromium-Nickel Stainless Steel Plate, Sheet and Strip for Fusion-Welded Unfired Pressure Vessels[2]
A 441 Specification for High-Strength Low-Alloy Structural Manganese Vanadium Steel[2]
A 572 Specification for High-Strength Low-Alloy Columbium-Vanadium Steels of Structural Quality[2]
D 1143 Testing Piles Under Static Axial Compressive Load[3]
D 3689 Testing Individual Piles Under Static Axial Tensile Load[3]

2.2 *ANSI Standards:*[4]
B30.1 Safety Code for Jacks
B46.1 Surface Texture

3. Significance and Use

3.1 The actual lateral load capacity of the pile-soil system can best be determined by lateral testing. Such testing measures the response of the pile-soil system to lateral loads and may

---

[1] This method is under the jurisdiction of ASTM Committee D-18 on Soil and Rock and is the direct responsibility of Subcommittee D18.11 on Deep Foundations.
Current edition approved March 2, 1981. Published May 1981.
[2] *Annual Book of ASTM Standards*, Part 4.
[3] *Annual Book of ASTM Standards*, Part 19.
[4] Available from the American National Standards Institute, 1430 Broadway, New York, N. Y. 10018.

provide data for research and development, engineering design, quality control, and acceptance or rejection under specifications.

3.2 Under the iterative elastic method of analysis that considers the nonlinear response of the soil[5], lateral testing combined with proper instrumentation can be used to determine soil properties necessary for the structural design of the pile to resist the lateral load to be applied.

3.3 Lateral testing as covered herein, when combined with an acceptance criterion, is suitable for control of pile foundation design and installation under building codes, standards, and other regulatory statutes.

4. Apparatus for Applying Loads

4.1 *General:*

4.1.1 The apparatus for applying lateral loads to the test pile(s) shall be as described in 4.3 or 4.4, unless otherwise specified, and shall be constructed so that the resultant loads are applied horizontally and in line with the central vertical axis of the pile or pile group so as to minimize eccentric loading and avoid a vertical load component.

NOTE 1—For lateral tests on batter pile frames or pile groups involving batter piles, consideration should be given to applying the lateral test loads at the actual or theoretical point of intersection of the longitudinal axis of the piles in the frame or group.

4.1.2 The test area within a radius of 20 ft (6 m) from the test pile or group shall be excavated or filled to the final grade elevation before testing the pile or pile group. If necessary, the pile(s) shall be cut off or extended so that the pile butt(s) is sufficiently above adjacent ground surface to permit construction of the load application apparatus, placement of the necessary testing and instrumentation equipment, and observation of the instrumentation. Before applying the test load, any annular space around the upper portion of the test pile(s) should be filled with sand or other suitable material and the same material and backfilling methods should be used for all production piles. Lateral test loads shall be applied at approximately pile cut-off elevation.

4.1.3 For tests on pile groups, except batter pile frames, the group of piles shall be capped with a reinforced concrete cap designed and constructed in accordance with accepted engineering practice to distribute the test loads uniformly to the piles in the group or shall be interconnected with steel members designed and constructed so that the piles act together. The connection between piles and pile caps and the depth of embedment of the pile butts into the pile cap shall simulate in-service conditions. Pile caps shall be cast above grade unless otherwise specified and may be formed on the ground surface unless 4.1.3.1 is specified.

4.1.3.1 *Elimination of Friction Beneath Pile Cap* (Optional)—For tests on pile groups, the bottom of the pile cap shall be clear of the ground surface.

NOTE 2—It is recommended that the bottom of the pile cap be clear of the ground surface when the friction between the soil and the pile cap may contribute significantly to the lateral resistance of the pile group.

4.1.3.2 *Passive Soil Pressures Against Pile Cap* (Optional)—For tests on pile groups, the pile cap shall be constructed below ground surface and backfilled with compacted fill on the side opposite the point of load application or the pile cap shall be constructed above ground surface against an embankment sufficient to permit the passive soil pressures to act during the test. If specified, compacted fill shall be placed against the sides of the pile cap to the extent practicable.

4.1.4 A steel test plate(s) of sufficient stiffness to prevent it from bending under the involved loads, but not less than 2 in. (50 mm) thick, shall be set vertically against the side of the pile, pile cap, or steel frame at the point(s) of load application and perpendicular to the line(s) of load application. The test plate shall be of sufficient size to accommodate the hydraulic cylinders but shall have a horizontal side dimension not less than one half the diameter or side dimension of the test pile(s) nor greater than the diameter or side dimension of the test pile(s). For tests on single piles other than square piles, the head of the pile shall be capped so as to provide a plane vertical bearing surface for the test plate or, the test plate shall be set in high-strength grout or adequately welded to the side of the pile using suitable filler material to provide full bearing against the projected area of the pile. If the test plate(s) is supported independently of the test pile or

---

[5] Reece, L. C., "Design and Evaluation of Load Tests on Deep Foundations," *Behavior of Deep Foundations, ASTM STP 670*, Am. Soc. Testing Mats., 1979.

group during assembly of the testing apparatus, such temporary supports shall be removed when test loads are applied.

4.1.5 *Bearing Plates*, shall be of steel and of sufficient size to accommodate spherical bearings, load cells, hydraulic jacks, and struts, and to transmit the applied lateral loads without detrimental high unit pressures. Bearing plates shall be of adequate thickness to prevent bending under the applied load but shall not be less than 2 in. (50 mm) thick.

4.1.6 *Struts and Blocking*—Struts shall be of steel and of sufficient size and stiffness to transmit the applied test loads without bending or buckling. Blocking used between reaction piles or between the hydraulic cylinder and the reaction system shall be of sufficient size and strength to prevent crushing or other distortion under the applied test loads.

4.2 *Testing Equipment:*

4.2.1 Unless the test load is applied by pulling in accordance with 4.5, lateral loads shall be applied using one or more hydraulic cylinders equipped with spherical bearings. If two or more hydraulic cylinders are to be used to apply the test load, they shall be of the same piston diameter, connected to a common manifold and pressure gage, and operated by a single hydraulic pump.

4.2.2 Hydraulic jacks including their operation shall conform to the applicable provisions of ANSI B30.1.

4.2.3 Unless a calibrated load cell(s) or equivalent device(s) is used, the complete jacking system including the hydraulic cylinder(s), valves, hydraulic pump, and pressure gage shall be calibrated as a unit to an accuracy of not less than 5 % of the applied load.

4.2.4 When an accuracy greater than that obtainable with the jacking system is required, a properly constructed load cell(s) or equivalent device(s) shall be used in series with the hydraulic cylinder(s). Load cells or equivalent devices shall be calibrated to an accuracy of not less than 2 % of the applied load and shall be equipped with spherical bearings.

4.2.5 If the lateral load is applied by pulling (4.5.4) the equipment used to produce the pulling force shall be capable of applying steady constant forces over the required load testing range. The dynamometer(s) or other in-line load indicating device(s) shall be calibrated to an accuracy of not less than 10 % of the applied load.

4.2.6 Calibration of testing equipment shall be done before each test or series of tests in a test program. Hydraulic cylinders shall be calibrated by loading the test equipment with the hydraulic cylinders over their complete range of piston travel for increasing and decreasing applied loads. Double-acting hydraulic cylinders shall be calibrated in both the push and pull modes. Calibration reports shall be furnished for all testing equipment for which calibration is required and shall show the temperature at which the calibration was done.

4.3 *Load Applied by Hydraulic Jack(s) Acting Against a Reaction System* (Fig. 1):

4.3.1 *General*—Apply the test loads to the pile or pile group using one or more hydraulic cylinders and a suitable reaction system according to 4.3.2, 4.3.3, 4.3.4, or 4.3.5. The reaction system may be any convenient distance from the test pile or pile group and shall provide a resistance greater than the anticipated maximum lateral test load. Set the hydraulic cylinder(s) (with load cell(s) if used) against the test plate(s) at the point(s) of load application in a horizontal position and on the line(s) of load application. Place a steel strut(s) or suitable blocking between the base(s) of the cylinder(s) and the reaction system with steel bearing plates in accordance with 4.1.5 between the strut(s) or blocking and the cylinder(s) and between the strut(s) and the reaction system. If a steel strut(s) is used, place it horizontally and on the line(s) of load application and brace the strut(s) to ensure it does not shift during load application. If two hydraulic cylinders are used, place both cylinders, load cells (if used), and struts or blocking at the same level and equidistant from a line parallel to the lines of load application and passing through the center of the test group. Support the jack(s), bearing plate(s), strut(s), and blocking on cribbing if necessary for stability.

4.3.2 *Reaction Piles* (Fig. 1a)—Install two or more reaction piles vertically or on a batter (or a combination of vertical and batter) so as to provide the necessary reactive capacity for the maximum anticipated lateral test loads. Cap the reaction piles with reinforced concrete, steel, or timber, or brace between the piles, or fasten the pile butts together so as to develop the lateral resistance of the entire group.

NOTE 3—Unless two opposing batter reaction piles are installed, the batter piles should be battered in a direction away from the test pile or group (see Fig. 1a).

4.3.3 *Deadman* (Fig. 1b)—Where soil or site conditions are suitable, install a deadman consisting of cribbing, timber panels, sheeting, or similar construction bearing against an embankment or the sides of an excavation so as to provide the necessary reactive capacity to the maximum anticipated lateral test loads.

4.3.4 *Weighted Platforms* (Fig. 1c)—Construct a platform of any suitable material such as timber, concrete, or steel, and load the platform with sufficient weights to provide the necessary resistance to the maximum anticipated lateral test loads to be applied. Provide a suitable bearing surface on the edge of the platform against which the reactive lateral load will be applied.

4.3.5 *Other Reaction Systems* (Optional)—Use any other specified suitable reaction system such as an existing structure.

4.4 *Load Applied by Hydraulic Jack(s) Acting Between Two Test Piles or Test Pile Groups* (Fig. 2)—Test the lateral capacity of two single piles or two similar pile groups simultaneously by applying either a compressive or tensile force between the pile or pile groups with a hydraulic jack(s). Test piles or test groups may be any convenient distance apart. If necessary, insert a steel strut(s) between the hydraulic cylinder(s) and one of the test piles or groups. For the cylinder(s), load cell(s) (if used), strut(s), and bearing plate(s) (if used), comply with the requirements of 4.1.5, 4.1.6, and 4.3.1, except remove all temporary blocking and cribbing underneath plates, strut(s), and cylinder(s) (and load cell(s) if used), after the first load increment has been applied and do not brace the strut(s).

4.5 *Load Applied by Pulling* (Optional):

4.5.1 *General*—Apply the lateral load by pulling test pile or group using a suitable power source such as a hydraulic jack, turnbuckle or winch connected to the test pile or group with a suitable tension member such as a wire rope or a steel rod and connected to an adequate reaction system or anchorage. Securely fasten the tension member to the test pile or pile cap so that the line of load application passes through the vertical central axis of the test pile or group. If two tension members are used, fasten them to the test pile or pile cap at points equidistant from a line parallel to the lines of load application and passing through the vertical central axis of the test pile or group.

4.5.2 *Anchorage System*—Maintain a clear distance of not less than 20 ft (6 m) or 20 pile diameters between the test pile or group and the reaction or anchorage system complying with 4.3.2, 4.3.3, 4.3.4, 4.3.5, or as otherwise specified. Furnish an anchorage system sufficient to resist without significant movement the reaction to the maximum lateral load to be applied to the test pile or group.

4.5.3 *Pulling Load Applied By Hydraulic Jack Acting Against a Reaction System* (Fig. 3)—Apply the lateral tensile load to the test pile or pile group using any suitable hydraulic cylinder such as conventional type, push-pull type, or center-hole type. Center the conventional hydraulic cylinder, (and load cell if used), on the line of load application with its base bearing against a suitable reaction system and its piston acting against a suitable yoke attached by means of two parallel tension members to the test pile or pile group (see Fig. 3a). Where required to adequately transmit the jacking load, install steel bearing plates in accordance with 4.1.5. If a double-acting type cylinder is used (Fig. 3b), place the cylinder on the line of load application connecting the cylinder's casing to the anchorage system and its piston to a suitable strut or steel rod adequately secured to the test pile or pile group. The steel strut or rod may be supported at intermediate points provided such supports do not restrain the strut or rod from moving in the direction of load application. If a center-hole cylinder is used (Fig. 3c), center the cylinder, (and load cell if used), along the line of load application with its base bearing against a suitable reaction and with its piston acting against a suitable clamp or nut attached to a steel rod or cable fastened securely to the test pile or group. Provide a hole through the reaction system for the tension member. If necessary to transmit the jacking forces, insert a steel bearing plate in accordance with 4.1.5 between the reaction and the jack base.

4.5.4 *Pulling Load Applied By Other Power Source Acting Against An Anchorage System*

(Fig. 4)—Apply the lateral tensile load with a winch or other suitable device. Insert a dynamometer or other load indicating device in the pulling line between the power source and the test pile or group (see Fig. 4a). If a multiple part line is used, insert the dynamometer or equivalent device in the line connecting the pulling blocks with either the test pile (or group) or the anchorage system. (See Fig. 4b).

4.6 *Fixed-Head Test* (Optional):

4.6.1 *Individual Pile* (Fig. 5)—Install the test pile so that it extends a sufficient distance above the adjacent ground surface to accommodate the steel frames but not less than 6½ ft (2 m). Firmly attach by clamping, welding, or some other means, a right angle (approximately 30-60-90) frame to each side of that portion of the pile extending above ground surface. Design and construct the frame so as to prevent the top of the pile from rotating under the maximum lateral load to be applied. Support the ends of the frames on steel rollers acting between steel bearing plates with the bottom bearing plate supported on a pile(s) or cribbing with sufficient bearing capacity to prevent any significant vertical deflections of the ends of the frame. Maintain a clear distance of not less than 10 ft (3 m) between the test pile and support for the ends of the frames. The steel bearing plate shall be of sufficient size to accommodate the ends of the frames and the steel rollers including the maximum anticipated lateral travel. Steel rollers shall be solid and shall be of sufficient number and diameter (but not less than 2 in. (50 mm) in diameter) so as to permit free horizontal movement of the frames under the anticipated downward pressures resulting from the maximum lateral test load to be applied.

NOTE 4—For practical purposes for a 10-ft (3-m) spacing between the test pile and frame support, it can be assumed that the vertical reaction at the ends of the frames is equal to the lateral load being applied to the test pile at the ground surface.

4.6.2 *Pile Group* (Fig. 6)—Install the test piles with pile tops a sufficient distance above the point of load application to provide fixity when the test group is capped. Cap the test group with an adequately designed and constructed reinforced concrete or steel grillage cap with sufficient embedment of the piles in the cap to provide fixity and with the side of the cap opposite the point of load application extended a sufficient distance to provide for the support pile(s). To prevent rotation of the pile cap under lateral load, support the end of the cap opposite that of the point of load application on one or more bearing piles with steel plates and rollers in accordance with 4.6.1 between the bottom of the cap and the top of the bearing pile(s).

4.7 *Combined Lateral and Axial Loading* (Optional):

4.7.1 *General*—Test the pile or pile group under a combination of lateral loading and axial compressive or tensile loading as specified. Apply the lateral load using method 4.3 or 4.4. Employ suitable methods and construction to ensure that the pile or pile group is not significantly restrained from lateral movement by the axial load.

4.7.2 *Compressive Load* (Fig. 7)—Apply the specified axial compressive load in accordance with 3.3 or 3.4 of Method D 1143. Place an antifriction device in accordance with 4.7.2.1, 4.7.2.2, or as otherwise specified between the compressive loading jack and the test plate on top of the test pile or pile group.

4.7.2.1 *Plate and Roller Assembly* (Fig. 8a)—The plate and roller assembly shall be designed to support the maximum applied compressive load without crushing or flattening of rollers and without indention or distortion of plates, and to provide minimal restraint to the lateral movement of the test pile or group as the lateral test loads are applied. Figure 8a illustrates a typical assembly having a compressive load limit of 100 tons (890 kN). The two plates shall be of Specification A 441 steel or equal with a minimum yield strength of 42 000 psi (290 MPa) and shall have a minimum thickness of 3 in. (75 mm). The plates shall have sufficient lateral dimensions to accommodate the length of rollers required for the compressive loads and for the anticipated travel of the rollers as the test pile or group moves laterally under load. The contacting surfaces of the steel plates shall have a minimum surface roughness of 63 as defined and measured by ANSI B46.1. The rollers shall be of sufficient number and length to accommodate the compressive loads and shall be of Specification A 572 steel Grade 45 or equal (minimum yield strength 45 000 psi (310 MPa) with a minimum diameter of 3 ± 0.001 in. (75 ± 0.03 mm). The rollers shall have a minimum surface roughness of 63 as defined and measured by ANSI B46.1. The plates shall be set level and the rollers shall be placed perpendicular to the direction of lateral load application with adequate spacing to prevent binding as lateral movement occurs.

4.7.2.2 *Antifriction Plate Assembly* (Fig. 8b)—The antifriction plate assembly shall be designed and constructed as illustrated in Fig. 8b and shall consist of the following elements: (*1*) a minimum 1-in. (25-mm) thick steel plate, (*2*) a minimum 10-gage (3.4-mm) steel plate tack welded to the 1-in. thick plate, (*3*) a minimum 3/32-in. (2.4-mm) sheet of virgin tetrafluoroethylene polymer with reinforcing aggregates prebonded to the 10-gage plate by a heat-cured epoxy, and (*4*) a minimum 1/4-in (6.4-mm) thick plate of Specification A 240 Type 304 stainless steel having a minimum surface roughness of 4 as defined and measured by ANSI B46.1. The area of contact between the tetrafluoroethylene polymer and the stainless steel plate shall be sufficient to maintain a unit pressure of less than 2000 psi (14 MPa) under the compressive loads to be applied. The area of the stainless steel plate shall be sufficient to maintain full surface contact with the tetrafluoroethylene polymer as the test pile or group deflects laterally. The stainless steel plate shall be formed with lips on opposite sides to engage the edges of the test plate under the lateral load. During the lateral test, the lips shall be oriented in the direction of the applied lateral load. The use of a plate assembly having an equivalent sliding friction shall be permitted. The use of two steel plates with a layer of grease in between shall not be permitted.

NOTE 5—Combined lateral and axial compressive loading is recommended to simulate in-service conditions. Precautions should be taken to avoid a vertical component resulting from the applied lateral load or a lateral component from the applied axial load.

NOTE 6—An apparatus for applying an axial tensile load to the test pile in combination with a lateral test load is difficult to construct without restraining the test pile from moving laterally under the lateral test loads. If it is required that a pile be tested under combined axial tensile and lateral loading, the use of a suitable crane equipped with a line load indicator is suggested for applying the uplift or tensile loads. Some type of universal acting device should be used in the tension member connecting the test pile with the crane hook. That in combination with the crane falls, should minimize restraint against lateral movement of the test pile under lateral loads.

5. Apparatus for Measuring Movements

5.1 *General*:

5.1.1 Set all reference beams and wires level and support them independently with supports firmly embedded in the ground and at a clear distance of not less than 7 ft (2 m) from the test pile(s) or group. Reference beams shall be of sufficient axial and lateral rigidity to provide stable reference points for pile deflection measurements. If a steel reference beam is used, one end of the beam shall be free to move horizontally as the length of beam changes with temperature variations.

5.1.2 Dial gage stems shall have at least a 3-in. (75-mm) travel and sufficient gage blocks shall be provided to allow for the maximum anticipated travel. Gages shall have a precision of at least 0.01 in. (0.25 mm). Provide smooth bearing surfaces perpendicular to the direction of the gage stem travel for all gage stems. Scales used to measure movements shall read to 1/64 in. or 0.01 in. (0.25 mm). Target rods shall read to 0.001 ft (0.3 mm).

5.1.3 Clearly mark all dial gages, scales, and reference points with a reference number or letter to assist in recording data accurately. Protect the instrumentation measuring system and reference system from adverse temperature variations and from accidental disturbance. Mount all gages, scales, or reference points so as to prevent movement relative to their support system during the test.

5.2 *Pile Butt Movements*—The apparatus for measuring lateral movement and recovery of the test pile or group along the line of load application shall consist of a primary and secondary system in accordance with the following methods:

NOTE 7—Two separate measuring systems are required for determining lateral movements of the test pile or group in order to have a check on the observed data, to provide for accidental disturbance of the measuring system, and to permit continuity of data in case it becomes necessary to reset the gages or scales.

5.2.1 *Dial Gage*—Orient the reference beam perpendicular to the line of load application. If the reference beam is located on the side of the test pile or group opposite the point of load application, allow sufficient clearance between the test pile or pile cap and the reference beam for the anticipated lateral movement of the pile or pile group. Mount the gage(s) on the reference beam with stem(s) bearing against the side of the pile or pile cap or mount the gage(s) on lugs attached to the test pile or pile cap with stems bearing against the reference beam. Mount the gages so their stems are horizontal and for single piles, along the line of load application. For tests on pile groups, mount the dial gages equidistant from the central line of load application.

5.2.2 *Wire, Mirror, and Scale* (Fig. 9)—Mount a mirror and scale on the top center of the test pile or pile cap or on a bracket mounted along the line of load application on the side of the test pile or cap with the scale along the line of load application. Stretch a piano wire or equivalent type perpendicular to the line of load application and passing over the face of the scale. Locate the wire not more than 1 in. (25 mm) from the face of the scale and at the supports install a suitable device to maintain tension in the wire throughout the test so that when plucked or tapped, the wire will return to its original position. If the scale and wire is placed on the side of the pile or cap opposite the point of load application, allow sufficient clearance between the pile or cap and the wire to provide for anticipated lateral movements of the pile or group.

5.2.3 *Transit and Scale*—Mount a scale horizontally on the side or top of the test pile or pile cap parallel to the line of load application and readable from the side. Establish outside of the immediate test area a permanent transit station and a permanent backsight or foresight reference point on a line perpendicular to the line of load application and passing through the target scale. With an engineer's transit, take readings on the target scale of lateral movements of the test pile or group referenced to the fixed backsight or foresight.

5.2.4 *Other Types of Measuring Apparatus*—Any other type of measuring device such as electrical or optical gages that yield accuracy equivalent to 0.01 in. (0.25 mm) may be used.

5.3 *Rotational Movement* (Optional) (Fig. 10)—For a test on a single pile(s) measure the rotation of the head of the test pile(s). Firmly attach to or embed in the test pile(s) a steel extension member in axial alignment with the test pile(s) and extending a minimum of 2 ft (0.6 m). Mount a dial gage on a reference beam with the gage stem horizontal and on the line of load application and bearing against the side of the extension member near its top (Fig. 10a). For tests on pile groups, measure the rotation of the pile cap by either (*1*) readings on reference points on top of the pile cap on the line of load application and on opposite ends of the cap using either dial gages mounted on an independent reference system or with a surveyors level reading a target rod on the reference points or vertical scales mounted on the pile cap at the reference point and referenced to a fixed bench mark; (*2*) a dial gage mounted on a reference beam a minimum of 2 ft (0.6 m) vertically above the dial gage used to measure pile butt movements (5.2.1) with its stem horizontal and on the line of load application and bearing against the side of the pile cap or a suitable extension thereto (Fig. 10b). For fixed-head tests on individual piles, use the apparatus for measuring rotation of free-head tests except that the upper dial gage stem may bear against the pile or measure the vertical movements at the ends of the steel frames using either a dial gage or a surveyor's level with a target rod or scale (Fig. 10c).

5.4 *Vertical Movement* (Optional)—Measure the vertical movements of the test pile(s) or pile group in accordance with 4.2 of Method D 1143 except that only one measuring system shall be required. For a test on an individual pile(s), a single reference point on the pile(s) is sufficient and for a test on a pile group, take readings on two reference points on opposite sides of the pile cap and in line with the applied load.

5.5 *Side Movement* (Optional)—Measure the movement of the test pile(s) or pile group in a direction perpendicular to the line of load application using either a dial gage mounted on a reference beam with the gage stem bearing against the side of the pile or pile cap or a scale mounted horizontally on the pile or pile cap perpendicular to the line of load application and read with an engineer's transit set up at a fixed position with the line of sight referenced to a fixed foresight or backsight.

NOTE 8—The measurement of vertical and side movements of the test pile under lateral loading may reveal eccentric loading or an abnormal behavior of the test pile. Such measurements are recommended if the precise response of the test pile to the lateral test load is required.

5.6 *Movement of Testing Apparatus* (Optional):

5.6.1 *Lateral Longitudinal Movements*—Measure the movements along the line of load application of the reference beam(s) and reaction system as well as the crushing of reaction system members using either an engineer's transit reading target scales attached to the reference beam(s) and the reaction system at strategic locations along the line of load application or by using dial gages suitably mounted and referenced. For transit readings, establish permanent transit stations and fixed backsights or foresights outside of the immediate test area.

5.6.2 *Vertical Movement*—Measure vertical movements of the reference beam(s) and reaction system using a surveyors level reading a target rod or scale located at strategic reference points along the line of load application. Reference level readings to a fixed bench mark located outside of the test area.

NOTE 8—To improve the reliability of measurements of test pile movements under load, it is recommended that the lateral and vertical movements of reference beams and the reaction system be measured in accordance with 5.6.

5.7 *Axial Deflections* (Optional)—Install in or on the test pile(s) to the depth(s) specified, tubing or ducts suitable to accommodate the types of inclinometer specified to be used.

NOTE 9—Except for very short stiff piles, inclinometer measurements are generally not warranted for the full length of the pile. Generally such measurements can be limited to the upper ⅓ to ½ of the pile length. The entire instrumentation system including materials, installation, equipment, and use should be set forth in the project specifications to the extent that this work is the responsibility of the contractor conducting the load tests.

6. Loading Procedures

6.1 *Standard Loading Procedures*—Unless failure occurs first, apply and remove a total test load equal to 200 % of the proposed lateral design load of the pile or pile group as follows:

Standard Loading Schedule

Percentage of Design Load	Load Duration, min
0	...
25	10
50	10
75	15
100	20
125	20
150	20
170	20
180	20
190	20
200	60
150	10
100	10
50	10
0	...

NOTE 10—Consideration should be given to limiting the lateral test load to that which would produce a maximum specified lateral movement, established for safety and load stability reasons.

6.2 *Loading in Excess of Standard Test Load* (Optional)—After applying and removing the standard test load in accordance with 6.1 (and 6.3 for standard loading if applicable), apply and remove the additional specified test loads in accordance with the following table:

Excess Loading Schedule
(following 6.1 loading)

Percentage of Design Load	Load Duration, min
0	10
50	10
100	10
150	10
200	10
210	15
220	15
230	15
240	15
250	15
etc. to maximum load specified in 10 % increments	etc. at 15 min intervals
max	30
75 max	10
50 max	10
25 max	10
0	...

6.3 *Cyclic Loading* (Optional)—Apply and remove the test load in accordance with the following table:

Cyclic Loading Schedules

Standard Loading

Percentage of Design Load	Load Duration, min	Percentage of Design Load	Load Duration, min
0	—	75	10
25	10	0	10
50	10	50	10
25	10	100	10
0	10	150	10
50	10	170	20
75	15	180	20
100	20	190	20
50	10	200	60
0	10	150	10
50	10	100	10
100	10	50	10
125	20	0	...
150	20		

Cyclic Loading Schedules

Excess Loading[A]			
Percentage of Design Load	Load Duration, min	Percentage of Design Load	Load Duration, min
Follow standard cylic loading schedule to 200 %		100	10
		0	10
		50	10
200	60	100	10
100	10	150	10
0	10	200	10
50	10	250	10
100	10	260	15
150	10	270	15
200	10	280	15
210	15	290	15
220	15	300	30
230	15	225	10
240	15	150	10
250	15	75	10
200	10	0	...

[A] Schedule for 300 % maximum load. For loading in excess of 300 %, hold 300 % load for 15 min, follow loading and holding time pattern for additional loading and hold maximum load for 30 min.

6.4 *Surge Loading* (Optional):

6.4.1 *General*—Surge loading involves the application of any specified number of multiple loading cycles at any specified load level. Surge loading may be applied in conjunction with standard loading or after the completion of standard loading. Apply surge loads at a uniform rate by continuous activation of the hydraulic jack (or other power source) and remove the surge load at a uniform rate by continuous release of the power source.

6.4.2 *Surge Loading with Standard Loading*—Apply and remove the test load in accordance with the following table:

Surge Loading Schedule[A] with Standard Loading

Percentage of Design Load	Load Duration, min
0	...
25	10
50	10
75	15
100	20
50	10
0	10
100	...
0	...
100	...
0	...
50	10
75	10
100	10
125	20
150	20
75	10
0	10
150	...
0	...
150	...
0	...
50	10
100	10
150	10
170	20
180	20
190	20
200	60
100	10
0	10
200	...
0	...
200	...
150	10
100	10
50	10
0	...

[A] Schedule shown for two surges at three load levels. If additional surges are specified or at other load levels follow the same loading and holding pattern.

6.4.3 *Surge Loading After Standard Load*—After applying and removing loads in accordance with 6.1, reapply the load to each specified load level and for the specified number of loading cycles, allowing sufficient time at each zero and peak load level for taking and recording the required load-movement data.

6.5 *Reverse Loading* (Optional)—Reverse loading involves the application of lateral test loads in either the push mode followed by the pull mode or vice versa. Test the pile or pile group in accordance with the loading schedule in 6.1, 6.2, 6.3, or 6.4 as specified first in one direction and then in the opposite direction.

6.6 *Reciprocal Loading* (Optional)—Apply and remove each specified lateral load level first in one direction and then in the opposite direction for the number of specified cycles. Hold each peak and zero load until load-deflection readings can be taken.

NOTE 11—Suitable apparatus is required to permit reversing the loads. Double-acting hydraulic cylinders are available in various sizes that can be activated by hand-operated, electric-powered, or air-hydraulic-powered pumps. Figure 11 illustrates various possible setups for applying reverse and reciprocal loading. Reciprocal loads can be applied with a suitable powered crank and connecting rod system combined with a device to measure the applied loads.

6.7 *Loading to Specified Total Lateral Movement* (Optional)—Apply the lateral test loads in accordance with 6.1, 6.2, 6.3, or 6.4 as specified until the gross lateral movement of the test pile or group is as specified and then remove the test load in four equal decrements allowing ten minutes between decrements.

6.8 *Combined Loading*—When the pile or pile group is tested under combined loading, in accordance with 4.7, apply the specified axial load before applying the lateral loads and hold the axial load constant during the application of the lateral loads in accordance with 6.1, 6.2, 6.3, or 6.4, as specified.

7. Procedures for Measuring Movements

7.1 *General*—Take required readings at each properly identified gage, scale, or reference point as nearly simultaneously as practicable. Clearly indicate and explain any adjustments made during the tests to the instrumentation or to the data recorded in the field. Also clearly explain any discontinuities in the data. If method 5.2.2 is used, take readings by lining up the wire with its image in the mirror.

7.2 *Standard Measuring Procedures*—Take and record readings of time, load, and movement immediately before and after the application of each load increment and the removal of each load decrement. Take and record additional readings at 5-min intervals between load increments and load decrements. While the total test load is applied, take and record readings at not less than 15-min intervals. Take and record readings 15 min and 30 min after the total load has been removed. If pile failure occurs, take the reading immediately before removing the first load decrement.

7.3 *Measurements for Surge Loading*—For initial application of test loads, for holding periods, for initial removal of the load and after removal of all loads, take and record the readings of time, load, and movement in accordance with 7.2. For the surge loading, take and record readings at the start and end of each load application.

7.4 *Measurements for Combined Loading*—If load tests are conducted in accordance with 4.7, take and record readings of vertical and side movements of the test pile(s) or group in accordance with 5.4 and 5.5 before and after the axial load is applied and removed.

7.5 *Measurements for Rotational Movements*—If observation of rotational movements of the test pile is specified (5.3), take and record the readings of rotational movement immediately before and after the application of each load increment and the removal of each load decrement. Take and record final recovery readings 30 min after the total test load has been removed.

7.6 *Measurements for Fixed-Head Tests, for Vertical and Side Movements, and for Movements of Testing Apparatus*—If the requirements of 4.6, 5.4, 5.5, or 5.6 are specified, take and record the readings before any test load is applied, at the proposed design load, at the maximum applied load and after all loads have been removed. Intermediate readings may be required if such measurements during testing appear unusual.

8. Safety Requirements

8.1 All operations in connection with pile load testing should be carried out in such a manner so as to minimize, avoid, or eliminate the exposure of people to hazards. Following are examples of safety rules to be followed in addition to general safety requirements applicable to construction operations.

8.1.1 Keep all work areas, walkways, and platforms, clear of scrap, debris, and small tools, and accumulations of snow and ice, mud, grease, oil, or other slippery substances.

8.1.2 All timbers and blocking and cribbing material shall be of quality material and be in good serviceable condition with flat surfaces and without rounded edges.

8.1.3 Hydraulic cylinders shall be equipped with spherical bearing plates, shall be in complete and firm contact with the bearing surfaces, and shall be aligned so as to avoid eccentric loading.

8.1.4 All struts used to transfer test loads to the reaction system or to another test pile or group shall be of steel, and shall be of sufficient size, strength, and stiffness to resist without excessive bending or deflection, a compression load 25 % greater than the maximum test load to be applied.

8.1.5 All tension rods used for pull tests shall be of sufficient size and strength to resist without excessive elongation, a tension load 25 % greater than the maximum test load to be applied and shall be adequately connected to the test pile or group, to the hydraulic cylinder and to the anchorage system.

8.1.6 All lines, rope, and cable, used for pull

tests shall be in good serviceable condition, free of abrasive wear, broken strands, kinks, and knots, and shall be of sufficient strength to resist a load 50 % greater than the maximum test load to be applied and shall be adequately connected to the test pile or group, to the power source, and to the reaction system.

8.1.7 All reaction systems shall be designed and constructed to have a reactive capacity sufficient to resist a load 25 % greater than the maximum test load to be applied.

8.1.8 All struts, blocking, bearing plates, and testing equipment shall be accurately aligned to minimize eccentric loading, and where necessary shall be restrained from shifting as test loads are applied so as not to affect the test results.

8.1.9 Attachments to the test pile(s), pile cap, or reaction system shall be designed and installed to transmit the required loads with an adequate factor of safety.

8.1.10 Loads shall not be hoisted, swung, or suspended over anyone and shall be controlled by tag lines.

8.1.11 All personnel shall stand clear of the jacking or pulling systems whenever test loads are being applied.

8.1.12 Only authorized personnel shall be permitted within the immediate test area.

9. Report

9.1 The report of the load test shall include the following information when applicable:

9.1.1 *General*:
9.1.1.1 Project identification,
9.1.1.2 Project location,
9.1.1.3 Test site location,
9.1.1.4 Owner,
9.1.1.5 Structural (foundation) engineer,
9.1.1.6 Geotechnical engineer,
9.1.1.7 Pile contractor,
9.1.1.8 Test boring contractor,
9.1.1.9 Designation and location of nearest test boring with reference to test pile or group,
9.1.1.10 Log of nearest test boring,
9.1.1.11 Horizontal control datum, and
9.1.1.12 Vertical control (elevation) datum.

9.1.2 *Pile Installation Equipment*:
9.1.2.1 Make, model, type, and size of hammer,
9.1.2.2 Rated energy of hammer, and
9.1.2.3 Size of predrilling or jetting equipment.

9.1.3 *Test and Reaction Piles*:
9.1.3.1 Identification of test and reaction piles,
9.1.3.2 Location of test piles and anchorage or reaction system,
9.1.3.3 Design load of pile or pile group,
9.1.3.4 Type of pile(s)—test and reaction,
9.1.3.5 Test pile material including basic specifications,
9.1.3.6 Tip and butt dimensions of pile(s),
9.1.3.7 General quality of timber test piles including occurrence of knots, splits, checks, and shakes, and straightness of piles,
9.1.3.8 Preservative treatment and conditioning process used for timber test piles including inspection certificates,
9.1.3.9 Wall thickness of pipe test pile,
9.1.3.10 Weight per foot of H-test pile,
9.1.3.11 Description of banding—timber piles,
9.1.3.12 Date precast test piles made,
9.1.3.13 Concrete cylinder strengths when pile tested (approximate),
9.1.3.14 Description of internal reinforcement used in test pile (size, length, number, longitudinal bars, arrangement, spiral or tie steel),
9.1.3.15 Condition of precast piles including spalled areas, cracks, head surface, and straightness of piles,
9.1.3.16 Effective prestress,
9.1.3.17 Number of piles in test or reaction group,
9.1.3.18 Which piles vertical or batter,
9.1.3.19 Degree of batter,
9.1.3.20 Embedded length—test and reaction piles,
9.1.3.21 Final elevation of test pile butt(s) and the ground surface at test pile, referenced to fixed datum, and
9.1.3.22 The depth of excavation and the distance from test pile(s) to adjacent excavation banks.

9.1.4 *Pile Installation-Test and Reaction*:
9.1.4.1 Date driven (installed),
9.1.4.2 Date concreted (cast-in-place),
9.1.4.3 Description of concrete (grout) mix including slump (cast-in-place),
9.1.4.4 Volume of concrete or grout placed in pile,
9.1.4.5 Description of pre-excavation or jetting (depth, size, pressure, duration),
9.1.4.6 Description of special installation

procedures used,

9.1.4.7 Type and location of pile splices,

9.1.4.8 Driving logs (blows per foot),

9.1.4.9 Final penetration resistance (blows per inch),

9.1.4.10 Cause and duration of interruptions in pile installation, and

9.1.4.11 Notations of any unusual occurrences during installation.

9.1.5 *Pile Testing*:

9.1.5.1 Date tested,

9.1.5.2 Type of lateral test,

9.1.5.3 Brief description of load application apparatus, including jack capacity,

9.1.5.4 Description of instrumentation used to measure pile movement including location of gages or other reference points (see Note 13),

9.1.5.5 Description of special instrumentation such as inclinometers,

9.1.5.6 Point of load application with reference to top of pile or pile cap, and to ground surface.

9.1.5.7 Special testing procedures used,

9.1.5.8 Axial load—type, amount, how applied,

9.1.5.9 Identification and location sketch of all gages, scales, and reference points (see Note 13),

9.1.5.10 Tabulation of all time, load, and movement readings,

9.1.5.11 Tabulation of inclinometer readings, declination versus depth,

9.1.5.12 Description and explanation of adjustments made to instrumentation, or field, data, or both,

9.1.5.13 Notation of any unusual occurrences during testing,

9.1.5.14 Test jack and other required calibration reports, and

9.1.5.15 Temperature and weather conditions during tests.

NOTE 12—In addition to the above required information to be reported, the results of any in-place and laboratory soil tests should be made available for the proper evaluation of test results.

NOTE 13—Suitable photographs can be very helpful in showing the instrumentation set-up, location of gages, scales, and reference points:

10. Precision and Accuracy

10.1 Inasmuch as each pile load test is unique, its results cannot be considered representative of another test unless similar in all respects; nor can the results be considered representative of a repeat test on the same pile. Therefore a general precision and accuracy statement is not warranted.

10.2 The required precision and accuracy of selected testing equipment and instrumentation has been included in this standard.

The American Society for Testing and Materials takes no position respecting the validity of any patent rights asserted in connection with any item mentioned in this standard. Users of this standard are expressly advised that determination of the validity of any such patent rights, and the risk of infringement of such rights, are entirely their own responsibility.

This standard is subject to revision at any time by the responsible technical committee and must be reviewed every five years and if not revised, either reapproved or withdrawn. Your comments are invited either for revision of this standard or for additional standards and should be addressed to ASTM Headquarters. Your comments will receive careful consideration at a meeting of the responsible technical committee, which you may attend. If you feel that your comments have not received a fair hearing you should make your views known to the ASTM Committee on Standards, 1916 Race St., Philadelphia, Pa. 19103, which will schedule a further hearing regarding your comments. Failing satisfaction there, you may appeal to the ASTM Board of Directors.

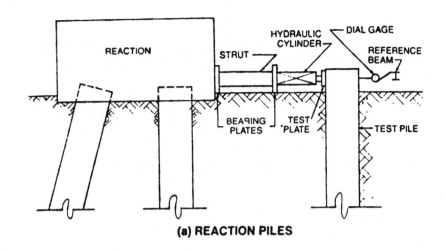

(a) REACTION PILES

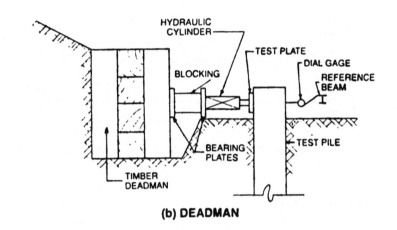

(b) DEADMAN

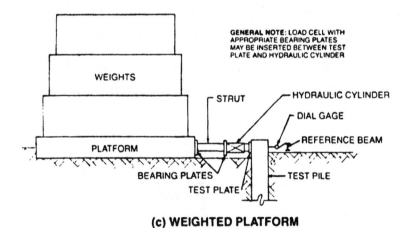

(c) WEIGHTED PLATFORM

FIG. 1 Typical Set Ups for Applying Lateral Load with Conventional Hydraulic Jack

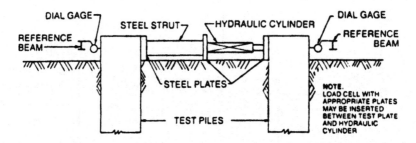

FIG. 2 Typical Arrangement for Testing Two Piles Simultaneously

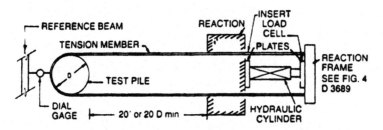

(a) CONVENTIONAL HYDRAULIC CYLINDER

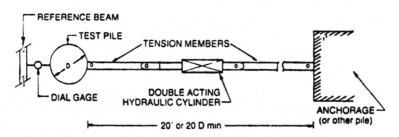

(b) DOUBLE ACTING HYDRAULIC CYLINDER

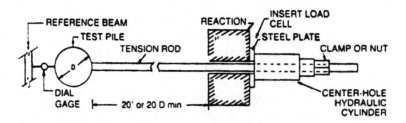

(c) CENTER-HOLE HYDRAULIC CYLINDER

FIG. 3 Typical Arrangements for Applying Pulling Loads with Hydraulic Jack (Top Views)

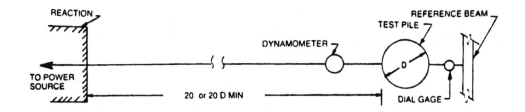

(a) SINGLE LINE

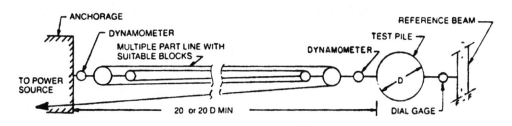

(b) MULTIPLE PART LINE

FIG. 4 Typical Arrangements for Applying Lateral Loads with Power Source such as Winch (Top Views)

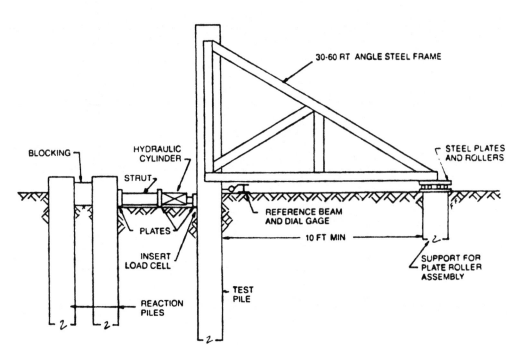

FIG. 5 Example of Fixed-Head Test Set Up for Lateral Test on Individual Pile

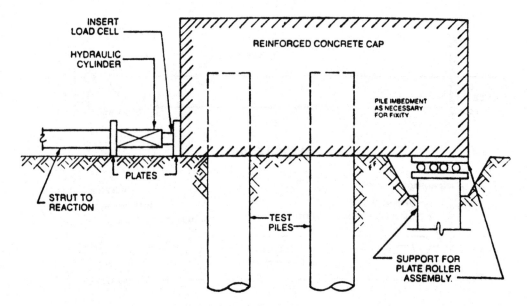

FIG. 6 Example of Fixed-Head Test Set Up for Lateral Test on Pile Group

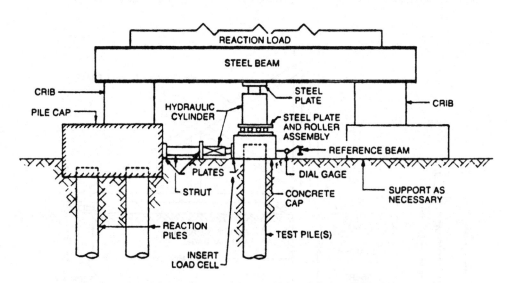

FIG. 7 Typical Example of Set Up For Combined Lateral and Axial Compressive Load

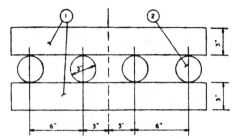

① ASTM A 441 PLATES WITH MINIMUM SURFACE ROUGHNESS OF 63 PER ANSI B 46.1
② FOUR ASTM A 572 GRADE 45 BARS 3 + 0.001 IN. DIAMETER BY 22 IN. LONG WITH MINIMUM ROUGHNESS OF 63 PER ANSI B 46.1

NOTE: LOAD LIMIT 100 TONS

(a) PLATE AND ROLLER ASSEMBLY

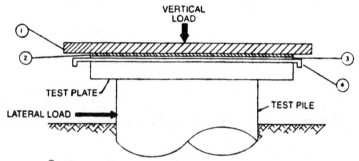

① ASTM A 36 PLATE MINIMUM 1 IN. THICK
② ASTM A 36 PLATE 10 G TACK WELDED TO ①
③ VIRGIN TETRAFLUOROETHYLENE POLYMER WITH REINFORCING AGGREGATES 3/32 IN. THICK PREBONDED TO ② WITH HEAT CURED EPOXY (FOR EXAMPLE FLUOROGOLD SLIDE BEARING¹)
④ ASTM A 240 TYPE 304 STAINLESS STEEL PLATE 1/4 IN. THICK FORMED AS SHOWN AND WITH MINIMUM SURFACE ROUGHNESS OF 4 PER ANSI B 46.1

NOTE: LOAD LIMIT 2000 P.S.I.

1. THE FLUOROCARBON COMPANY 317 CHANGE BRIDGE ROAD PINE BROOK N.J. 07058

(b) ANTIFRICTION PLATE ASSEMBLY

FIG. 8 Typical Antifriction Devices for Combined Load Test

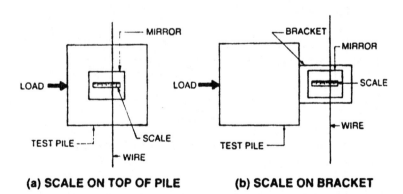

(a) SCALE ON TOP OF PILE   **(b) SCALE ON BRACKET**

FIG. 9 Typical Wire-Scale Arrangements to Measure Lateral Deflections (Top Views)

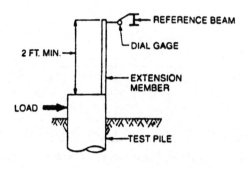

(a) SINGLE PILE

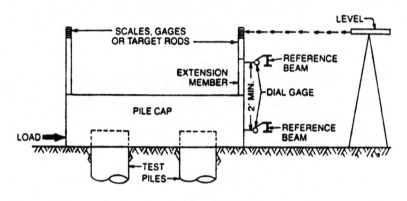

(b) PILE GROUP

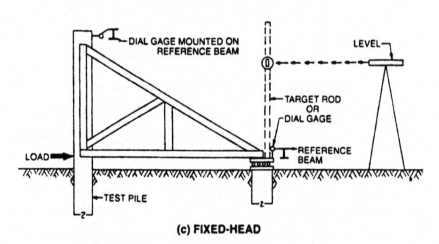

(c) FIXED-HEAD

FIG. 10 Typical Arrangements for Measuring Pile Head Rotation

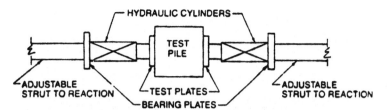

NOTES:
1. USE SINGLE-ACTING SOLID-PLUNGER HYDRAULIC CYLINDER (PUSH TYPE)
2. RELEASE CONNECTION BETWEEN TEST PILE AND REACTION ON SIDE OPPOSITE THAT OF LOAD APPLICATION (E G REMOVE JACK BLOCKING, STRUT ETC OR RELEASE ADJUSTABLE STRUT)
3. PREVENT RESTRAINTS AGAINST LATERAL MOVEMENT OF TEST PILE UNDER LOAD
4. NOT SUITABLE FOR RECIPROCAL LOADING
5. LOAD CELLS WITH APPROPRIATE PLATES COULD BE INSERTED BETWEEN HYDRAULIC CYLINDERS AND TEST PLATES

(a) WITH STANDARD HYDRAULIC CYLINDER

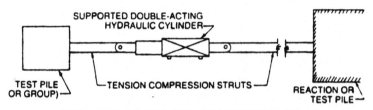

NOTES:
1. USE DOUBLE ACTING HYDRAULIC CYLINDER IN COMBINATION WITH 4 WAY DIRECTIONAL VALVE
2. START TEST WITH PLUNGER EXTENDED TO ONE HALF ITS STROKE
3. LATERAL MOVEMENT OF TEST PILE(S) EITHER SIDE OF CENTER IS LIMITED TO 1 2 THE PLUNGER STROKE
4. STROKES RANGE FROM 8 TO 14 INCHES
5. SUITABLE FOR RECIPROCAL LOADING

(b) WITH DOUBLE-ACTING HYDRAULIC CYLINDER

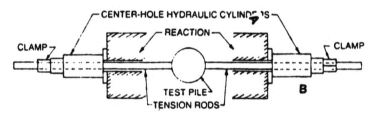

NOTES
1. USE SINGLE-ACTING HOLLOW-PLUNGER HYDRAULIC CYLINDERS IN COMBINATION WITH 4 WAY DIRECTIONAL VALVE
2. LATERAL MOVEMENT OF TEST PILE EITHER SIDE OF CENTER IS LIMITED TO 1 2 PLUNGER STROKE OR FROM 1 TO 1-1 2 INCHES UNLESS SYSTEM OPERATED AS FOLLOWS
   A RELEASE CLAMP A – ACTIVATE CYLINDER B
   B RELEASE LOAD FROM CYLINDER B
   C RELEASE CLAMP B – ACTIVATE CYLINDER A
   D RELEASE LOAD FROM CYLINDER A
3. SUITABLE FOR LIMITED RECIPROCAL LOADING
4. LOAD CELLS WITH APPROPRIATE PLATES COULD BE INSERTED BETWEEN REACTIONS AND HYDRAULIC CYLINDERS

(c) WITH TWO CENTER-HOLE HYDRAULIC CYLINDERS

FIG. 11 Typical Reverse Lateral Loading Set Ups

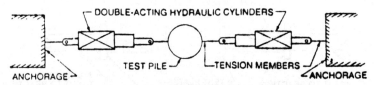

NOTES
1. USE DOUBLE-ACTING HYDRAULIC CYLINDERS IN THE PULL MODE IN COMBINATION WITH A 4 WAY DIRECTIONAL VALVE
2. LATERAL MOVEMENT OF TEST PILE EITHER SIDE OF CENTER IS LIMITED TO THE PLUNGER STROKE WHICH RANGES FROM 8 TO 14 INCHES
3. SUITABLE FOR RECIPROCAL LOADING
4. DYNAMOMETER IN TENSION LINES CAN BE USED TO MEASURE LOAD

(d) WITH TWO PULL-TYPE HYDRAULIC CYLINDERS

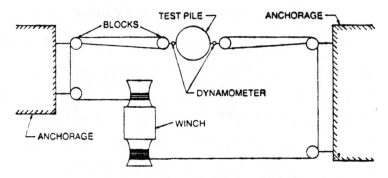

(e) WITH TWO-DRUM WINCH

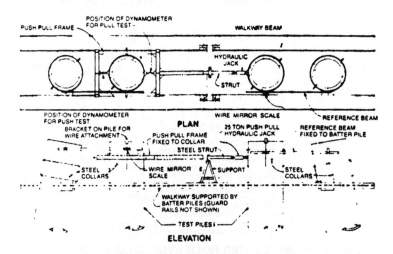

(f) WITH SPECIAL PUSH-PULL FRAME

(COURTESY RAYMOND INTERNATIONAL INC.)

FIG. 11—*continued*

REFERENCES

REFERENCES

Awoshika, Katsuyuki, and Reese, L. C., "Analysis of Foundation with Widely Spaced Batter Piles," Research Report 117-3F, Project 3-5-68-117, Center for Highway Research, The University of Texas at Austin, February, 1971.

Banerjee, P. K. and Davies, T. G., "Analysis of Some Reported Case Histories of Laterally Loaded Pile Groups," Proceedings, Numerical Methods in Offshore Piling, The Institution of Civil Engineers, London, May, 1979, pp. 101-108.

Bogard, D., and Matlock, Hudson, "Procedures for Analysis of Laterally Loaded Pile Groups in Soft Clay," Proceedings, Geotechnical Practice in Offshore Engineering, American Society of Civil Engineers, April, 1983.

Broms, Bengt B., "Lateral Resistance of Piles in Cohesive Soils," Proceedings, American Society of Civil Engineers, Vol. 90, No. SM2, March, 1964a, pp. 27-63.

Broms, Bengt B., "Lateral Resistance of Piles in Cohesionless Soils," Proceedings, American Society of Civil Engineers, Vol. 90, No. SM3, May, 1964b, pp. 123-156.

Broms, Bengt B., "Design of Laterally Loaded Piles," Proceedings, American Society of Civil Engineers, Vol. 91, No. SM3, May, 1965, pp. 79-99.

Bryant, L. M. "Three-Dimensional Analysis of Framed Structures with Nonlinear Pile Foundations," Unpublished Dissertation, The University of Texas at Austin, 95 pages, December, 1977.

Canadian Foundation Engineering Manual, Part 3, Deep Foundations, March, 1978, p. 98.

Cappozolli, L., "Test Pile Program at St. Gabriel, Louisiana," Louis J. Capozzoli and Associates, July, 1968.

Cox, W. R., Reese, L. C., and Grubbs, B. R., "Field Testing of Laterally Loaded Piles in Sand," Proceedings, Offshore Technology Conference, Paper 2079, Houston, Texas, May, 1974.

Coyle, H. M., and Reese, L. C., "Load Transfer for Axially Loaded Piles in Clay," Proceedings, American Society of Civil Engineers, Vol. 92, No. SM2, March, 1966, pp. 1-26.

Coyle, H. M., and Sulaiman, I. H., "Skin Friction for Steel Piles in Sand," Proceedings, American Society for Civil Engineers, Vol. 93, No. SM6, November, 1967, pp. 261-278.

Fenske, Carl W., Personal communication, 1981.

Gibbs, H. J., and Holtz, W. G., "Research on Determining the Density of Sands by Spoon Penetration Testing," *Proceedings*, Fourth International Conference on Soil Mechanics and Foundation Engineering, London, Vol. 1, 1957, pp. 35-39.

Hargrove, J.Q., "Field Test of a Pile Under Lateral Loading in a Layered Soil Profile," Unpublished Thesis, The University of Texas at Austin, 127 pages, May, 1981.

Hetenyi, M., *Beams on Elastic Foundation*, The University of Michigan Press, Ann Arbor, 1946.

Horne, Michael, *Plastic Theory of Structures*, Pergamon Press, New York, 1978.

Japanese Road Association, "Road Bridge Substructure Design Guide and Explanatory Notes, Designing of Pile Foundation," May, 1976, p. 67 (in English).

Kraft, L. M., Jr., Ray, R. P., and Kagawa, T., "Theoretical t-z Curves," *Proceedings*, American Society of Civil Engineers, Vol. 107, No. GT11, November, 1981, pp. 1543-1561.

Kubo, K., "Experimental Study of the Behavior of Laterally Loaded Piles," *Proceedings*, Sixth International Conference on Soil Mechanics and Foundation Engineers, Vol. II, Montreal, 1965, pp. 275-279.

Kuthy, R. A., Ungerer, R. P., Renfrew, W. W., Hiss, J. G. F., Jr., and Rizzuto, I. F., "Lateral Load Capacity of Vertical Pile Groups," Research Report 47, Engineering Research and Development Bureau, New York State Department of Transportation, Albany, April, 1977, 37 pages.

Lam, Philip, "Computer Program of Analysis of Widely Spaced Batter Piles," Unpublished Thesis, The University of Texas at Austin, August, 1981.

Lowe, Gerald, and Reese, L. C., "Analysis of Drilled-Shaft Foundations for Overhead-Sign Structures," Center for Transportation Research Report 244-2F, The University of Texas at Austin, May, 1982.

Matlock, Hudson, "Correlations for Design of Laterally Loaded Piles in Soft Clay," *Proceedings*, Second Annual Offshore Technology Conference, Paper No. OTC 1204, Houston, Texas, Vol. 1, 1970, pp. 577-594.

Matlock, Hudson, and Reese, L. C., "Foundation Analysis of Offshore Pile-Supported Structure," *Proceedings*, Fifth International Conference, International Society of Soil Mechanics and Foundation Engineering, Paris, Vol. 2, 1961, p. 91.

Matlock, Hudson, Ingram, W. B., Kelley, A. E., and Bogard, D., "Field Tests of the Lateral Load Behavior of Pile Groups in Soft Clay," *Proceedings*, Twelfth Annual Offshore Technology Conference, OTC 3871, Houston, Texas, May, 1980.

McClelland, Bramlette, "Design and Performances of Deep Foundations," <u>Proceedings</u>, Specialty Conference on Performance of Earth and Earth Supported Structures, Purdue University, Soil Mechanics and Foundations Division, American Society of Civil Engineers, June, 1972.

O'Neill, M. W., "Group Action in Offshore Piles," <u>Proceedings</u>, Geotechnical Practice in Offshore Engineering, American Society of Civil Engineers, April, 1983.

O'Neill, M. W., Ghazzaly, O. I., and Ha, H. B., "Analysis of Three-Dimensional Pile Groups with Nonlinear Soil Response and Pile-Soil-Pile Interaction," <u>Proceedings</u>, Offshore Technology Conference, Houston, Texas, Vol. II, Paper No. 2838, 1977, pp. 245-256.

Poulos, H. G., "Behavior of Laterally Loaded Piles: II - Pile Groups," <u>Proceedings</u>, American Society of Civil Engineers, Vol. 97, No. SM5, May, 1971b, pp. 733-751.

Prakash, S., "Behavior of Pile Groups Subjected to Lateral Loads," Unpublished Dissertation, University of Illinois, Urbana, 1962.

Reese, L. C., "Load versus Settlement for an Axially Loaded Pile," <u>Proceedings</u>, Part II, Symposium on Bearing Capacity of Piles, Central Building Research Institute, Roorkee, February, 1964, pp. 18-38.

Reese, L. C., "Analysis of a Bridge Foundation Supported by Batter Piles," <u>Proceedings</u>, Fourth Annual Symposium on Engineering Geology and Soil Engineering, Moscow, Idaho, April, 1966, pp. 61-73.

Reese, L. C., "Design and Evaluation of Load Tests on Deep Foundations," Behavior of Deep Foundations, Symposium sponsored by American Society for Testing and Materials, Committee D18 on Soil & Rock for Engineering Purposes, Boston, Massachusetts, June 25-30, 1978, pp. 4-26.

Reese, L. C., "Behavior of Piles and Pile Groups under Lateral Load," a manual prepared for U.S. Department of Transportation, Federal Highway Administration, Office of Research, Washington, D. C., 1983.

Reese, L. C., Cox, W. R., and Koop, F. D., "Analysis of Laterally Loaded Piles in Sand," <u>Proceedings</u>, Fifth Annual Offshore Technology Conference, Vol. II, Paper No. OTC 2080, Houston, Texas, 1974, pp. 473-485.

Reese, L. C., Cox, W. R., and Koop, F. D., "Field Testing and Analysis of Laterally Loaded Piles in Stiff Clay," <u>Proceedings</u>, Seventh Annual Offshore Technology Conference, Vol. II, Paper No. OTC 2312, Houston, Texas, 1975, pp. 672-690.

Reese, L. C., and Matlock, Hudson, "Nondimensional Solutions for Laterally Loaded Piles with Soil Modulus Assumed Proportional to Depth," <u>Proceedings</u>, Eighth Texas Conference on Soil Modulus and Foundation Engineering, Special Publication No. 29, Bureau of Engineering Research, The University of Texas, Austin, Texas, September, 1956.

Reese, L. C., and Matlock, Hudson, "Behavior of a Two-Dimensional Pile Group Under Inclined and Eccentric Loading," *Proceedings*, Offshore Exploration Conference, Long Beach, California, 1966, pp. 123-140.

Reese, L. C., O'Neill, M. W., and Smith, R. E., "Generalized Analysis of Pile Foundations," *Proceedings*, American Society of Civil Engineers, Vol. 96, No. SM1, January, 1970 ,pp. 235-250.

Reese, L. C., and Nyman, K. J., "Field Load Tests of Instrumented Drilled Shafts at Islamorada, Florida," a report to Girdler Foundation and Exploration Corporation, Clearwater, Florida, Bureau of Engineering Research, The University of Texas at Austin, February 28, 1978.

Reese, L. C., and Welch, R. C., "Lateral Loading of Deep Foundations in Stiff Clay," *Proceedings*, American Society of Civil Engineers, Vol. 101, No. GT7, February, 1975, pp. 633-649.

Schmertmann, John H., "Report on Development of a Keys Limerock Shear Test for Drilled Shaft Design," a report to Girdler Foundation and Exploration Corporation, Clearwater, Florida, 1977.

Sullivan, W. R., "Development and Evaluation of a Unified Method for the Analysis of Laterally Loaded Piles in Clay," Unpublished Thesis, The University of Texas at Austin, May, 1977.

Terzaghi, K., "Evaluation of Coefficients of Subgrade Reaction," *Geotechnique*, Vol. V, 1955, pp. 297-326.

Vesic, A. S., "Bending of Beams Resting on Isotropic Elastic Solid," *Proceedings*, American Society of Civil Engineers, Vol. 87, No. SM2, April, 1961a, pp. 35-53.

Vesic, A. S., "Beams of Elastic Subgrade and the Winkler's Hypothesis," *Proceedings*, Fifth International Conference on Soil Mechanics and Foundation Engineering, Vol. 1, 1961b, Paris, France, pp. 845-850.

Welch, R. C., and Reese, L. C., "Laterally Loaded Behavior of Drilled Shafts," Research Report No. 3-5-65-89, Center for Highway Research, The University of Texas at Austin, May, 1972.